Molecules

MOLECULES

P. W. Atkins

W. H. FREEMAN AND COMPANY

NEW YORK

Library of Congress Cataloging-in-Publication Data

Atkins, P. W. (Peter William), 1940–
    Molecules

    (Scientific American Library series; #21)
    Bibliography: p.
    Includes index.
        1. Molecular structure.   2. Molecules.
        I. Title.   II. Series: Scientific American Library
    series; no. 21.
        QD461.A85   1987   547.1   87-20681

ISBN 0-7167-6004-5 (pbk.)

Printed in the United States of America.

Sixth printing 1996, HC

CONTENTS

Preface   vii

Introduction   1

1 SIMPLE SUBSTANCES   13
Air   14
Water and Ammonia   23
Smog, Pollution, and Acid Rain   27

2 FUELS, FATS, AND SOAPS   33
Natural Gas and LPG   34
Gasoline and Coal   37
Alcohol and Vinegar   42
Fats and Oils   50
Butter and Margarine   57
Soaps and Detergents   61

3 SYNTHETIC AND NATURAL POLYMERS   65
Polymers and Plastics   66
Rubber   77
Polyesters and Acrylics   80
Nylon   84
Hair, Wool, and Silk   88
Sugar, Starch, and Cellulose   95

4 TASTE, SMELL, AND PAIN 105
Sweetness 106
Sourness and Bitterness 110
Hot, Spicy, and Cool 114
Meatiness and Barbecues 117
Fruits and Foods 124
Flowers and Essential Oils 132
Animal Smells 140

5 SIGHT AND COLOR 145
Vision 146
Leaves, Carrots, and Flamingos 149
Flowers, Fruit, and Wine 155
Browns, Bruises, and Tans 158

6 THE LIGHT AND THE DARK 161
Pain Killers and Tranquilizers 162
Stimulants 167
Nasty Compounds 171
Sex 175

Glossary 179

Further Reading 183

Sources of the Illustrations 184

Index 187

PREFACE

Joy may be inarticulate, but reflection is empty without understanding. There is delight to be had merely by looking at the world, but that delight can be deepened when the mind's eye can penetrate the surface of things to see the connections within.

The following pages are intended to augment our delight when looking at the world. They introduce one facet of chemistry—its portrayal of the structure of substances—and they aim to show what makes up the things that make up the everyday world. These pages are an introduction to *molecules*. They are meant, among other things, to show the shapes of molecules and the connections between them, to show why some molecules behave as they do, and to reveal the economy of nature. In short, this book aims to make the molecular familiar.

I have purposely included molecules that I find interesting. Even so, I could have chosen any of a thousand others and still remained with the familiar. But limitations of space dictate the brevity of the selection, and I must ask readers to quell the irritation they will certainly feel when they look in the index for a substance and find it ignored. There are several million known compounds, and manufacturers of pills, potions, or detergents—and nature especially—have at their disposal a vast chemical organ on which they can conjure symphonies of mixtures. It is inevitable that a book such as this will be incomplete. Its purpose is only to open an eye, not to show the world.

There is no particular order in which the book should be read; indeed, it is not necessarily meant to be read in any order: It is a book for occasional delectation. I wrote it, however, with a particular flow of thought in mind, so it is not completely amorphous and can profitably be read from front to back.

I tried to avoid technical terms throughout, but some inevitably (or at least unintentionally) crept in and are explained in the Glossary. Where possible, I also tried to explain. But do not expect too much fulfillment here, for some explanations are not yet known to anyone, and others require too much technical background. Moreover, I did not want to diminish delight by overloading the text with too much explanation: This is only an *introduction* to understanding. I particularly wanted to show that *some* appreciation of the features of molecules can be achieved without a college degree (or even a freshman course) in chemistry.

Most of the information presented here has been culled from a dozen or so books that I have come to respect. They are listed under Further Reading, and I hope that if the authors find their thoughts on these pages, they will regard that as a tribute. Many of the points I mention are discussed in more detail in those books, and readers would be well-advised to check with them before using the information I provide—for I have also cut corners in my wish to simplify and render palatable a sometimes tough and complex dish.

I should like to thank all those who either wittingly or unwittingly have contributed to this book. It is always invidious to mention only some members of the team by name, leaving unsung all those others who have contributed so much. To the latter I am especially grateful, for the production of this book has been demanding. Among the former I would like to thank particularly Travis Amos, who researched the photographs and in doing so taught me a mode of thought. My heartfelt thanks also go to the team that Linda Chaput has assembled at W. H. Freeman and Company to produce the Scientific American Library series, especially to Andrew Kudlacik, Barbara Brooks, Linda B. Davis, and Mike Suh, who offered advice at the early stages; to Lloyd Black and Philip McCaffrey, who increasingly absorbed my energies as editing and production filled the sky; to Bill Page, who coordinated the illustration program; and to Lynn Pieroni, who brought the elements together into a beautiful page. My thanks go as well to Julia DeRosa, who had to oversee production on a very tight schedule.

As to matters technical, I should like to thank O. C. Dermer, Professor Emeritus, Oklahoma State University and Dr. A. J. MacDermott of the University of Oxford, who read and commented helpfully on the entire draft. I am also grateful to Dr. S. A. Greenfield of the University of Oxford for helping to clarify my thoughts on neurophysiology.

P. W. ATKINS
Oxford, 1987

Molecules

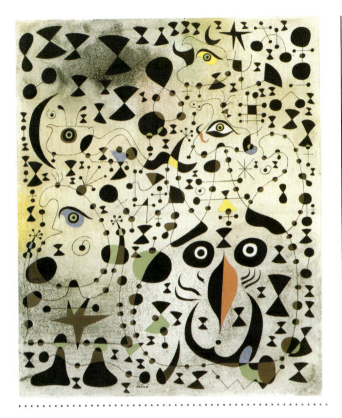

The Beautiful Bird Revealing the Unknown to a Pair of Lovers, 1941, by Joan Miró.

INTRODUCTION

When you hold this book you are holding molecules. When you drink coffee you are ingesting molecules. As you sit in a room you are bombarded by a continuous storm of molecules. When you admire the color of an orchid and the textures of a landscape you are admiring molecules. When you savor food and drink you are enjoying molecules. When you smell decay you are smelling molecules. You are clothed in molecules, you eat them, and you excrete them. In fact, you are made of molecules.

Molecules are characteristic groupings of atoms, of the kind shown in the drawings throughout this book. (I will define a molecule more precisely later.) We are surrounded by molecules, and most of what we touch is made of them. Yet many people do not know they exist. In fact, until the beginning of this century molecules were regarded as little more than symbols used by chemists to describe transformations of matter, and until recently no one had ever knowingly seen one.

In an extraordinary collaboration at the turn of this century, physicists, chemists, and biologists revealed the reality of molecules. Through *spectroscopy,* the study of the light emitted and absorbed by molecules, they were able to deduce the arrangement of atoms in individual molecules. Through *x-ray diffraction,* the study of the patterns in the intensity of x-rays when they pass through single crystals of substances, they were able to make out the concentrations of electrons corresponding to the atoms in a molecule. Extraordinary developments in microscopy, especially in electron microscopy, finally yielded what had long been sought—images of individual molecules and atoms. Now scientists can see more directly the shapes of molecules and begin to unravel the reasons matter behaves as it does.

The following pages are intended to show a little of what has been found. They will show the molecules we breathe, wear, eat, burn, and in general what constitutes the everyday. Their primary purpose is to acquaint you with the variety of molecules with a minimum of technical prerequisites. The structures of molecules can be apprehended directly without a deep knowledge of the sophisticated techniques involved in their determination. Molecules can be seen as bundles of atoms of different kinds locked together into a definite arrangement, and to imagine what it is you are wearing, eating, and smelling there is no need to know how chemists, nature, or industry contrive their manufacture. Nor is their any great need, until your interest rises, to know how each one carries out its function. The molecules are here to feed your imagination.

The molecules that are described range from the simplest possible to the highly complex. Some do ostensibly

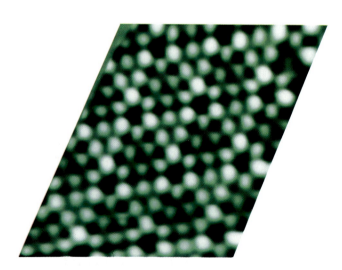

Modern techniques of electron microscopy have such good resolution that individual atoms can be seen. Here we see atoms of silicon on the surface of a sample of pure silica.

humdrum things, such as methane (16), which is merely burned. (Numbers in parentheses indicate molecules that are described in the text.) Others are included because they act as molecular building blocks, or they happen to typify a taste or an odor, or they are responsible for a color. Some molecules do very grand things and are included here because of their importance. Among these is the most ubiquitous chemical in the world, cellulose (85), which grows as great forests and softens the face of the earth. You will see how one or two atoms can convert a fuel to a poison, change a color, render an inedible substance edible, or replace a pungent odor with a fragrant one. That changing a single atom can have such consequences is the wonder of the chemical world.

The drawings alone should tell you much about the composition and appearances of the molecules, and your perusal could, with profit, stop there. I very much do not want you to believe that to comprehend the structure of the world it is necessary for you to graduate in or even to study chemistry. I hope you will comprehend the kingdom I will show with immediacy, without prerequisites, without labor, with a pleasure akin to looking at a work of art.

However, the drawings are enriched by knowing what they represent and understanding how a molecule performs its function. The remaining paragraphs of this introduction explain some of the background to the drawings, suggest how to think about them, and sketch a few of the arguments that lead from atoms to molecules to properties.

One widely used introductory chemistry text is subtitled *The Central Science*. This brilliant epithet summarizes beautifully why the following remarks are so important. Chemistry stands at the pivot of science. On the one hand it deals with biology and provides explanations for the processes of life. On the other hand it mingles with physics and finds explanations for chemical phenomena in the fundamental processes and particles of the universe. Chemistry links the familiar with the fundamental. These pages will give but a mere glimpse of this link, but, with luck, you will see a little through the mist and understand how a chemist thinks.

A single atom can make a considerable difference to the properties of a molecule. The molecules responsible for the blue of a cornflower and the red of a poppy differ by only one hydrogen atom. This is explained in more detail on page 156.

ELEMENTS AND ATOMS

One of the great achievements of chemistry was to show that all the matter in the world, be it a lump of rock, a glass of water, an ostrich feather, or a tree, is built from no more than about one hundred *elements*. The elements include hydrogen, carbon, oxygen, and copper, for example, and are so called because they cannot be broken down into simpler substances by heating, roasting, boiling, treatment with acid, or any of the other techniques that chemists use for changing matter. Physicists, of course, have developed more aggressive techniques, and they can smash elements apart into electrons, protons, and the other fundamental particles of nature using special high-energy particle accelerators. However, for our purpose, which is to explore our surroundings, we can stay with the one hundred elements and marvel that the rich tapestry of the world can be stitched from so meager a selection of thread.

The smallest amount of an element that can exist is an *atom* (from the Greek *atomos,* "uncuttable"). A lump of a pure element, such as a lump of pure gold, is a collection of identical atoms. An atom itself consists of a central positively charged *nucleus,* surrounded by one or more diffuse shells of negatively charged electrons that cancel its charge. Thus it is convenient to think of atoms as minute spheres, with each element having a characteristic radius.

The radius of a carbon atom is only about 1.5×10^{-10} meters (0.00000000015 meter) so that a 3-cen-

Five of the chemical elements. Clockwise from the yellow sulfur are copper, bromine, mercury, and iodine.

Niagara that the heat sears and consumes. The world and everything in it is built from the almost negligible.

Despite the minuteness of atoms, modern microscopes can make them out. Even newer techniques can be used virtually to feel atoms and make out their shapes. In other words, atoms are very small but real. In this book, they will be represented by spheres magnified some 25 million times, so that a carbon atom will look about 1 centimeter in diameter. Oxygen and nitrogen atoms have about the same number of electrons as carbon (8 and 7, respectively, in place of carbon's 6) and are almost the same size as carbon atoms. A hydrogen atom is appreciably smaller because in place of carbon's six electrons it has only one. Most of the other atoms that we will meet are appreciably larger than carbon. Phosphorus, sulfur, and chlorine atoms all have more than twice as many electrons as carbon (15, 16, and 17, respectively) and their diameters are about 50 percent greater. Nevertheless, whenever we are thinking of atoms, we are thinking of a world so small as to be almost, but not quite, beyond imagination.

timeter line of carbon (about this long: _____) is a hundred million carbon atoms from end to end and about a million atoms across. One of the wonders of this world is that objects so small can have such consequences: Any visible lump of matter—even the merest speck—contains more atoms than there are stars in our galaxy. Each type of atom brings a particular quality to every substance of which it is a part, and although atoms are so small, their colossal number in any tangible sample results in our perception of their properties. When we lift an apple we feel the weight of a galaxy of almost weightless atoms. When we hear the ripple of water we are hearing shockwaves as a myriad of almost imperceptible molecules crash down and collide with other molecules. When we dress we pull across our bodies a great web spun from almost infinitesimal dots and held together by the conspiracy of forces between them. When we see a flame we are seeing the release of an almost negligible droplet of energy, but in such a

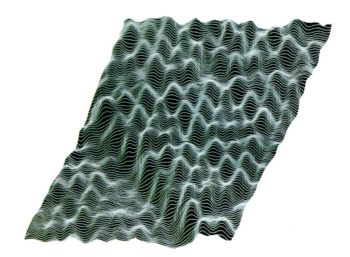

Surface tunneling microscopy is a new technique for revealing the surface structure of solids. In this image of the surface of silicon, the cliff running across the surface is one atom high.

COMPOUNDS

A *compound* is a combination of elements. Thus, water (6) is a combination of hydrogen and oxygen, and aspirin (146) is a combination of carbon, hydrogen, and oxygen.

Many compounds consist of *molecules*. A molecule, as mentioned before, is a specific, discrete grouping of atoms into a definite geometrical arrangement. (Numerous illustrations of models of molecules appear later in the book.) Almost all the molecules that are described here consist of no more than half a dozen elements, although some may include several atoms of each element. Because so few elements will be shown, it is convenient to distinguish them by spheres of different colors. The convention shown in the illustration below will be used throughout. Extraordinary as it may seem, from this palette of eight colors, we will build almost all the 160 compounds illustrated in these pages. Just occasionally, another element will be needed and will be introduced as required. Gray spheres will normally represent the atoms of metals (specifically, sodium and potassium).

The colors of the spheres are not the actual colors of the atoms. Atoms are colorless, diffuse, misty clouds of electrons surrounding a minute speck of a central nucleus, and the color of a molecule is a collective property of all its atoms (how we perceive colors is described under some entries, particularly 1 and 136). The colors used for the spheres are simply coding and are chosen because they allude to some property of the element or its compounds. Thus hydrogen is shown as white because it is the simplest atom; carbon is as black as soot; and oxygen, the life-giver, is red. Chlorine is admittedly a greenish-yellow gas (hence, its name, from the Greek *khloros,* "green") but its color is actually due to the chlorine molecule (a pair of chlorine atoms linked together), and not to each individual atom. Likewise, sulfur is a yellow element, but its color is a property of the eight-atom rings that are sulfur molecules.

BONDS BETWEEN ATOMS

Atoms do not stick together in arbitrary numbers and spatial arrangements. Instead, the atoms of a particular element will combine in only certain ways with the atoms of other elements (including, perhaps, its own atoms). The reason for the restriction is to be found in the properties of electrons and their arrangement around each atomic nucleus.

For our purposes it is necessary to know only that a link or *bond* between two atoms in a molecule is due to their sharing of a pair of electrons. (This idea, which was first proposed by the American chemist G. N. Lewis in the opening decades of this century, has survived the rigors of quantum mechanical scrutiny with only minor changes of detail.) The two electrons hover between the atoms and act as a kind of electrostatic glue between the nuclei. That is, *each bond in a molecule is a shared electron*

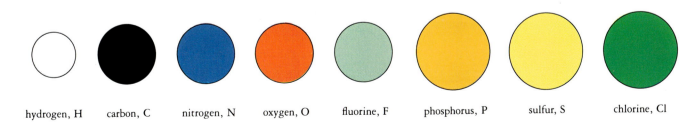

hydrogen, H   carbon, C   nitrogen, N   oxygen, O   fluorine, F   phosphorus, P   sulfur, S   chlorine, Cl

The atoms of the elements shown in the illustrations later in the book are distinguished by this color code.

pair. The number of bonds that an atom can form with other atoms is then a reflection of the number of electrons it can share with its neighbors. The rules governing this activity (which can be explained by going further into atomic structure) are as follows:

- a hydrogen atom usually forms only one bond
- a carbon atom usually forms four bonds
- a nitrogen atom usually forms three bonds
- an oxygen atom usually forms two bonds
- a chlorine atom usually forms one bond

Bond sharing is shown in the illustration on the opposite page. Hydrogen can also take part in a special type of bonding *between* molecules, in which it links two atoms; this is explained later.

In writing the structure of a molecule (as distinct from showing a model of it), it is common to represent a bond with a short single line (—) between the *chemical symbols* of the atoms it joins. The symbols we need include:

H	hydrogen
C	carbon
N	nitrogen
O	oxygen
F	fluorine
P	phosphorus
S	sulfur
Cl	chlorine

The bond between a hydrogen atom and a chlorine atom in hydrogen chloride is therefore represented as H—Cl.

Some atoms can form more than one bond to another atom. If a carbon atom shares two pairs of electrons with a neighboring oxygen atom, for example, then there is a *double bond* between them. This double bond is denoted $C=O$, and it can be seen in many of the structures shown later, including acetic acid (32) and testosterone (159). Similarly, a pair of atoms can share three pairs of electrons. Then the two atoms are joined by a *triple bond,* as in the hydrogen cyanide molecule H—C≡N (104), and the fabric Orlon (69). Atoms cannot share four pairs of electrons; hence, three bonds between two atoms is the maximum possible.

ORGANIC COMPOUNDS

Most of the compounds shown in the following pages are *organic.* That is, they are compounds containing carbon and (usually) hydrogen. Compounds that are not organic are called *inorganic.* Some very simple carbon compounds, particularly those not containing hydrogen (carbon dioxide, chalk and other carbonates, for example) are honorary inorganic compounds.

The term "organic" does not mean that the compounds are necessarily made by biological organisms, although that was once thought to be the case. It was once believed that organic compounds contained some kind of "vital force" that led to life. That view was dispelled in the nineteenth century, when it was shown that a typical organic compound, urea (130), a component of urine, could be made from inorganic starting materials. The petrochemical and pharmaceutical industries are based on the synthesis and modification of organic compounds.

Organic compounds are prominent in these pages because they are so important and, I find, interesting. They are responsible for the colors and odors of flowers and vegetation and for the taste of food. Indeed, virtually the whole of the natural world, apart from the rocks and the oceans, consists of organic compounds. Many of the newer construction materials, notably plastics, are also organic, as are almost all pharmaceuticals.

Carbon plays a special role in the world because it has a unique ability to form links with itself. A glance at the following pages will show many examples of molecules that consist of chains and rings of black carbon spheres. A few other elements can link to themselves, but none so

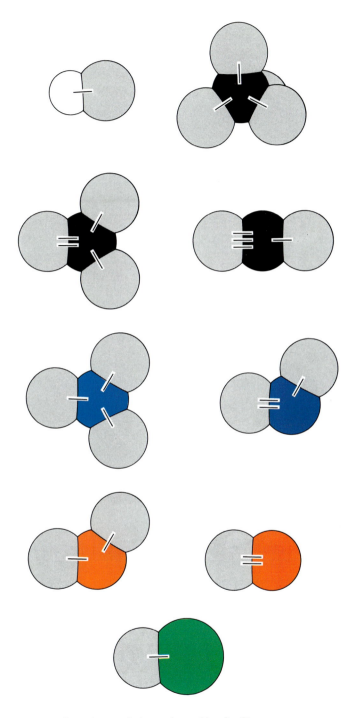

The atoms of the elements form characteristic numbers of bonds. The gray atoms represent atoms of elements that can form one, two, or three bonds.

extensively as carbon, and none gives so many stable structures. Carbon has this unique ability because it is rather mediocre as an element. It has a middling ability to attract electrons from other atoms, and its own electrons can be removed fairly easily. But what a versatile and wide-ranging mediocrity that is, for in it lies the potential for life.

In examining many of the organic compounds, it may be helpful to think of them as chains or rings of carbon atoms that form a carbon frame and to which are attached other groups of atoms. These *functional groups* are the chemically active parts of organic molecules. Examples include the carbon-carbon double bond ($\overset{\diagdown}{\diagup}C=C\overset{\diagup}{\diagdown}$), the *carbonyl group* ($\overset{\diagdown}{\diagup}C=O$), and the *hydroxyl group* (—O—H). A carbon-carbon double bond is normally a chemically sensitive part of a molecule, one that is liable to chemical attack. In the attack, one of the bonds bursts open and a group of atoms can attach to each of the carbon atoms it originally joined. A double bond is also responsible for holding a molecule into a rigid shape. Single bonds act like hinges, enabling molecules to fold into many different shapes, but a double bond is rigid and cannot be twisted. Hence a (not infallible) sign that a molecule is chemically active is its possession of double (or triple) carbon-carbon bonds, and a sign that a molecule is flexible and can roll up into a tight ball, or roll out into a straight line, is the absence of multiple bonds.

Many molecules, notably water (6) and ammonia (7), have atoms that possess pairs of electrons that are not involved in bonding. Such a *lone pair* can be pictured as two electrons concentrated in one region of the surface of an atom. A molecule with a lone pair is free to form a bond with any other atom that can share a pair of electrons. As a result, a lone pair can be thought of as a sticky patch on a molecule, to which other atoms may attach. A lone pair can also be thought of as a spearhead. A molecule can drive its lone pair into an atom of another molecule, perhaps forming a bond or even forcing out an atom already attached to the target molecule.

Hence, the possession of lone pairs is often a sign of chemical activity. It is one of the reasons ammonia is so pungent and water is such a good solvent for many materials.

STRUCTURES AND FORMULAS

The variety of structures that carbon can form is due in part to its unparalleled ability to form single, double, and triple bonds with another carbon atom or with other atoms. This structural fecundity allows it to form C—C, C=C, C≡C, C—O, C=O, C—N, C=N, and C≡N bonds and hence give rise to very intricate networks.

This intricacy presents the following problem. An interesting molecule, such as vanillin (126), may appear in a drawing to be little more than a random jumble of red, white, and black spheres. The eye needs more help than a drawing alone can give. For this reason, chemists often represent a molecular structure by a *line formula,* which shows only the links and virtually ignores the atoms. A few examples should make this clear. The benzene molecule (23) is a hexagonal arrangement of six carbon atoms and six hydrogen atoms. Its full structure is

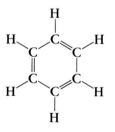

In the line formula, the hydrogen atoms are ignored, and only the carbon-carbon bonds are shown:

The related compound toluene (24),

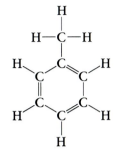

is shown as

with the single spike representing a CH$_3$ group. Atoms other than carbon are shown explicitly, as in octanoic acid:

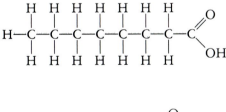

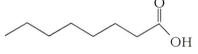

This convention greatly simplifies the drawing of organic structures, and it helps to unravel the arrangement of spheres in the model. (Some small groups, such as —CH$_3$ and —CHO, are often written explicitly for clarity.)

In what follows, a line formula is given whenever it is likely to be useful. Interpreting these formulas will seem easy after practicing with some structures and comparing them with the models. An example of this comparison is given in the illustrations below. On the upper left is the line formula. The first step needed to interpret it as a full structure is to replace each intersection and line end by a carbon atom, C, as shown on the upper right. Next, enough hydrogen atoms, H, are attached to each carbon atom to bring its total number of bonds to four. This produces the full structure at the bottom center.

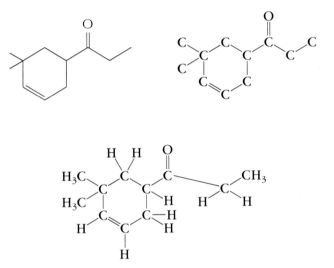

A final point on notation: It is sometimes helpful to specify the kinds of atoms that make up a molecule and the number of atoms of each type. This is done by writing the *molecular formula* of the compound, in which the symbol for each element is followed by a subscript showing the number of atoms of that element in the molecule. A trivial example is H$_2$O for water, showing that it consists of two hydrogen atoms and one oxygen atom. But a water molecule is a midget next to a molecule of the fat tristearin (36). Its molecular formula is C$_{57}$H$_{110}$O$_6$, showing that each tristearin molecule consists of 57 carbon atoms, 110 hydrogen atoms, and 6 oxygen atoms.

One technical point is that the same molecular formula may apply to two or more different substances. An example is the molecular formula C_2H_6O, which applies to both ethanol,

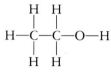

and to dimethyl ether,

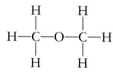

Compounds with the same molecular formula are called *isomers* of each other, from the Greek for "equal parts," and the different molecules can be thought of as built from the same kit of parts. A special kind of isomerism occurs when two molecules differ only in their shapes. This is called *geometrical isomerism* and is illustrated by the following two compounds:

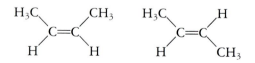

These are distinct compounds because the C=C double bond is rigid and one end of the molecule cannot be rotated relative to the other. In the text we will refer to the molecule pictured on the left as the cis isomer because the two CH_3 groups are on the same side of the molecule (and *cis* is Latin for "on this side"). The isomer pictured on the right is the trans isomer, because the CH_3 groups are on opposite sides (and *trans* is Latin for "across"). At this stage, isomerism may seem of minor technical importance. However, in due course (see molecule 158), we will see that it can have horrendous consequences.

NONMOLECULAR COMPOUNDS

Not all compounds are molecular. Another large class of inorganic compounds, for example, is *ionic*. The origin of the bonding in these compounds is the electrostatic attraction between oppositely charged ions.

An *ion* is an atom that has lost or gained one or more (negatively charged) electrons and hence has acquired an electric charge. When an atom loses electrons it becomes positively charged and is called a *cation*. An example is the sodium ion, Na^+, which is a sodium atom that has lost one electron. A copper cation, Cu^{2+}, is a copper atom that has lost two electrons. When an atom gains electrons it becomes negatively charged and is called an *anion*. A chlorine atom forms a chloride anion, Cl^-, by gaining one electron. An oxygen atom forms the oxide anion, O^{2-}, by gaining two electrons.

SOLIDS, LIQUIDS, AND GASES

A crystalline solid can be pictured as an orderly array of particles. The particles, which may be atoms in a solid element (copper atoms in copper), ions in an ionic compound (sodium cations and chloride anions in table salt), or molecules in a molecular compound (water molecules in ice), are almost stationary. They may vibrate and oscillate at their sites in the solid, but they are too tightly packed together to move past each other.

If the solid is heated, the particles vibrate more vigorously, and voids open up in the array. Now the particles can begin to move past each other, and the substance flows. The solid has melted to a liquid.

If the liquid is heated still more, the motion of the particles becomes more vigorous, and some can escape from its surface. These escaped, widely separated, rapidly moving particles have formed a gas.

THE VARIETIES OF SOLIDS

We need consider two principal classes of solid: a *molecular solid* and an *ionic solid*. A molecular solid consists of closely packed molecules. The molecules usually interact with each other only weakly, for all the bonding ability of their atoms has already been satisfied, and only some weak types of interaction remain. Because the molecules are held together so weakly, the solid is easily disrupted by gentle warmth, as in the case of fat. In some cases the molecules can also be pushed past each other by gentle mechanical pressure, which is why butter, a molecular solid, can be spread with a knife. Softness is often also a sign that the solid is a *mixture* of different but perhaps closely related molecules. In a soft solid, the different molecules have slightly different shapes and pack together only loosely. *Pure* molecular solids are often brittle because the molecules of a single compound pack together well. Ice, solid benzene, and sugar (81) are all brittle molecular compounds.

In some cases molecules do interact strongly with each other, and this adds to the rigidity of the solid. Water molecules interact strongly in ice (as do sucrose molecules in sugar) and bind themselves together into a rigid crystal. The origin of this strong interaction is a special kind of link called a *hydrogen bond*. Hydrogen bonds are so important and account for so many properties that they are worth describing in a little more detail.

A hydrogen bond, which is denoted by a dotted line in place of a single dash, consists of two atoms with a hydrogen atom lying between them. An example is A—O—H··O—B, where A—O—H and O—B are hypothetical generalized molecules. Such a bond seems to break the general rule about hydrogen forming only one bond. The bond is formed because the oxygen in A—O—H attracts electrons so strongly that it draws the shared electron pair in its own O—H bond toward itself, leaving the positive charge of the hydrogen nucleus almost fully exposed. That exposed positive charge is strongly attracted to other electrons nearby, particularly those of another oxygen atom, as in the molecule O—B. The hydrogen bond is therefore a vestigial ionic bond between two molecules. It can occur only between atoms that can attract electrons strongly, which for our limited palette of elements are

$$O—H\cdots O$$
$$O—H\cdots N$$
$$N—H\cdots O$$
$$N—H\cdots N$$

(Fluorine, F, is the only other element for which hydrogen bonding is important.) As you will see, hydrogen bonding is of enormous importance in the world, for among numerous other things, it accounts for the existence of oceans and the strength of wood.

Ionic solids are stacks of enormous numbers of cations and anions held together by their electrostatic attraction for each other. They tend to be hard, and they generally melt at high temperatures. Ionic solids form many of the

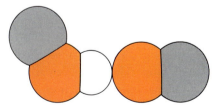

A hydrogen bond is formed by a hydrogen atom lying between the atoms of two strongly electron-attracting elements (typically oxygen). The gray spheres represent the rest of the molecule.

basic structures of the world, including the rocks, hills, and mountains of our landscapes and the bones of our skeletons. These solids are rigid—which accounts for the fact that we do not sink up to our knees when walking across the Rockies and, indeed, accounts for our knees as well—because the ions are bound together so strongly and cannot easily be pushed aside. Ionic solids are also often brittle, because a sharp blow can displace the cations from the anions, shattering the crystal.

MIXTURES

Most of the substances in the everyday world are mixtures of compounds or, in some cases, mixtures of elements. Most rocks and many minerals are complex mixtures of largely ionic compounds. Food and the living things that use landscapes as a stage (and which are often destined to become food themselves) are extremely complex mixtures of organic compounds, for they consist of varieties of biological cells, with all the paraphernalia of life packed inside. The flavor of an orange, for example, is due to an unconscious conspiracy of hundreds of different compounds.

The task of presenting the molecules of our natural and synthetic world might therefore seem hopeless. However, simplifications are possible. Some substances are single compounds and can be shown precisely, for example, aspirin (146) and sugar (81). For others, the molecules *typical* of their composition can be shown. This is the case with gasoline (page 37), a complex mixture that is dominated by several types of molecules. For still other substances, molecules that *typify* their characteristics can be shown. Thus, although there is not space here to show all the known components of an orange, typical molecules that contribute to its flavor or its texture can be illustrated. Hence, we can at least begin to distinguish an orange molecularly from a banana.

All this should go to show that the world is simultaneously simple and complex. It is complex because even a leaf is composed of myriad types of molecules with a composition that changes with the seasons. It is simple because scientists can unravel the leaf, distinguish its components, and see how one component may dominate in the summer but that another will dominate in the fall (page 157).

This knowledge about substances may be achieved by identifying the molecules and assembling information about their behavior. The tale of unraveling is far from complete, and it may be an age before we are certain that we know every detail about the differences between an orange and a banana, let alone a man and a woman. Nevertheless, these pages should show you that the horde of molecules that make up the world is becoming less an amorphous swarm and more a group of individual personalities.

Oranges have an odor that arises in part from a terpene (page 132) molecule (limonene) that is the mirror image of the molecule that contributes to the odor of lemons. The color of oranges is due largely to carotene molecules (136) but these blood oranges also include some anthocyanin molecules (139). Their tartness is due to citric acid (91).

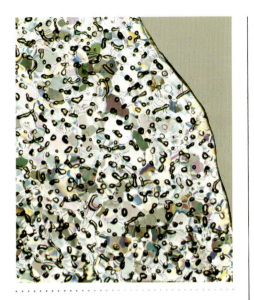

Ice with gas bubbles.

1

SIMPLE SUBSTANCES

Almost all the molecules discussed in this chapter are inorganic, and they are included here for two reasons. One is to provide the gentlest possible introduction to molecules. The other is to present some molecules that act as springboards for what is to come—as starting points for more complex molecules. The more complex molecules can be pictured as being formed by the insertion of an additional atom or group of atoms between two that are bonded together in the simpler molecule. Each new atom brings something of the personality of its element to the molecule, and this conspiracy of atoms results in a molecule with properties that are richer than those of each atom alone. The small molecules in this chapter show that even a minor modification, such as the insertion of a nitrogen atom between the two atoms of an oxygen molecule, can have profound consequences.

Small these molecules may be, but unimportant they are certainly not, for they include the gases of the air and the water of the oceans: They are the enablers, if not the components, of life. They are also agents of havoc, for they include the substances that pollute and destroy where nature, left alone, might flourish.

AIR

An early version of our atmosphere emerged from the earth beneath our feet, as at this outgassing at Isla Santa Cruz, the Galápagos, Ecuador.

When you stand in the open air, you are immersed in a gas that is about four-fifths nitrogen and one-fifth oxygen. (The illustrations that follow show the molecules present in dry, pollution-free air.) As in any gas, the molecules are in continuous chaotic motion (indeed, the word "gas" comes from the same Greek root as "chaos"). The molecules hurtle through space at about the speed of sound (about 700 miles per hour), collide with each other, and go hurtling off in other directions until they collide again a fraction of a second later. The incessant impact of this storm of molecules on the surface of a container—including your container, your skin—is experienced as a virtually constant pressure (about 14 pounds per square inch at sea level). On a still, warm summer's day, or in a quiet room such as you may now be in, you are in fact at the center of an unseen, unfelt storm of molecules. When the wind blows, the molecules stream predominantly in one direction and strike that side of your face.

Sometimes this unseen stream of molecules can be strong enough to fell trees and destroy buildings.

It is not at all clear where our atmosphere came from or how it has changed, although one guess about the latter is shown in the graph below. There is general agreement that an early atmosphere was formed as an *outgassing* of the rocks and planetesimals that aggregated to form our initially primitive planet. A similar kind of outgassing occurs today at volcanoes, and it is surmised that the gases they release—largely water vapor, hydrogen, hydrogen chloride, carbon monoxide, carbon dioxide, nitrogen, and molecules containing sulfur—were abundant in the first atmosphere. Of these, only nitrogen is abundant in our present atmosphere. Hence there is a question about where the rest have gone and where other gases have come from. Some answers—which are little more than wise (but perhaps false) speculations—are described here and in the next section. One substance can be dealt with at once: Hydrogen molecules, being very light and moving very fast, escaped the gravitational pull of the planet and disappeared into space, as any that are newly formed would soon do today.

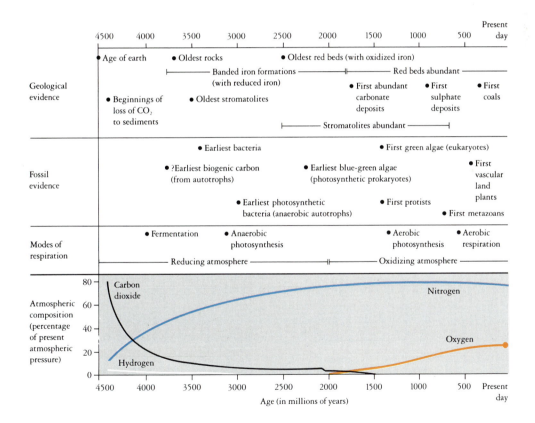

A speculative history of the atmosphere. Note the decline in carbon dioxide with the formation of sedimentary deposits and the correlation between the changes in composition and the biological and geological changes.

ARGON (1) Ar

Although argon is not the most abundant gas in the atmosphere, it is fitting that we begin with this simplest kind of substance: one that exists only as free atoms. Argon ("the unreactive one") is the third most abundant component of dry air, and each breath you take is about 1 percent argon. It is because argon is so unreactive that it exists in the air as atoms and forms no stable compounds.

Argon arrived from below the surface of the earth. An argon atom is formed when the nucleus of a potassium atom in the minerals of the earth captures one of the electrons that surrounds it and hence is *transmuted into argon. (Terms preceded by an asterisk are defined in the Glossary.) Unlike potassium, newly formed argon atoms cannot bind to the surrounding atoms, and they leak out of the ground as a gas. This process is still continuing and as the potassium dies and rocks decay the concentration of argon is slowly rising.

Argon is harvested from the atmosphere, which is our sole and barely sufficient source. Some is obtained from the *distillation of liquid air. Some also accumulates in the nitrogen used for the manufacture of ammonia (7) because, although the nitrogen combines with hydrogen in the process, the argon does not.

Large amounts of argon are used in the steel industry to dilute oxygen that is blown through the molten metal to burn away impurities. Some is used in incandescent light bulbs, where it helps to carry heat away from the tungsten filament without reacting with it. Some is used in fluorescent lamps, which are glass tubes filled with a mixture of argon and mercury vapor. The argon makes the lamps easier to start and helps to regulate the current through them. By a mechanism akin to that resulting in neon's red radiation (which is described later), argon also supplements the radiation provided by the *fluorescence of the phosphors (which include magnesium tungstate, zinc silicate, cadmium borate, and cadmium phosphate) that coat the inner side of the tube. These phosphors emit visible light when illuminated with the violet and ultraviolet light emitted by energetically excited mercury atoms.

Argon is one of the *noble gases,* a group of elements which includes helium (first detected on the sun and now collected from oil wells), neon ("the new one"), krypton ("the hidden one"), and xenon ("the stranger"). All exist as free atoms in small amounts in the atmosphere. The group also includes the radioactive gas radon, which bubbles up out of the ground as a product of nuclear disintegration within the earth's crust. Currently many people are concerned that radon outgassing from below buildings or from mineral-based construction materials may collect in health-threatening amounts in thermally insulated, and hence sealed, structures.

An electric discharge through argon and neon excites atoms to states of high energy; once excited they discard the excess energy as light.

Neon is familiar as a gas that fills advertising signs, for it emits a bright red glow when an electric current is passed through it. Spiced with a little mercury vapor, argon gives a blue discharge. Yellow and green signs are obtained by using neon and argon, respectively, in colored glass tubes. Neon-glow indicator lamps contain a mixture of argon and neon.

The origin of the colors is as follows: The electric current in the tube containing the gas is a storm of electrons. These electrons smash into atoms of gas and excite them to states of increased energy in which their own electrons are rearranged slightly. An excited atom shrugs off this extra energy almost immediately as its electrons collapse back into a lower energy arrangement, and the discarded energy is radiated away as light. The greater the amount of energy that needs to be lost, the shorter the *wavelength of the light. Low-energy transitions result in red light, intermediate in yellow, and high in blue. Very-high-energy transitions give rise to invisible, damaging, *ultraviolet light.*

The chemical inertness of argon and the other noble gases is due to their electronic structure (almost all explanations in chemistry can be traced to electrons and their arrangements around nuclei). In each case, the shells of electrons surrounding the nucleus are full and cannot accept any more electrons. Because noble gas atoms do not readily accept electrons (or lose them), they rarely enter into bond formation. Indeed, until the 1960s it was widely believed that the noble gases formed no compounds at all. They were then called the *inert gases.* However, chemists had to beat a hasty retreat when the first noble gas compounds (with the highly reactive gas fluorine) were formed; now several dozen compounds of krypton and xenon are known, most of which contain oxygen or fluorine. Helium, neon, and argon, however, are still chemical virgins.

....

OXYGEN   (2) O_2

Oxygen accounts for about 20 percent of the volume of the atmosphere. It is the most abundant element in the earth's crust, accounting for almost half its total mass and is present there in combination with other elements in the form of water, silicates, and oxides. It is also abundant on the moon, but not in the free state; there, oxygen is trapped in compounds in the lunar rocks. When space travel becomes more commonplace, we may find it economical to mine the lunar surface for its oxygen; here on earth we mine for metals and discard the oxygen from their ores.

Although oxygen is abundant in the atmosphere and is obtained industrially from this rich source by distillation of liquid air, it is a relative latecomer there. The atmosphere of the newly formed earth did not contain oxygen. Some of it arrived when water molecules (6) that had outgassed from the rocks were broken apart by the intense radiation from the sun, and the liberated oxygen atoms joined together in pairs. The bulk of our oxygen arrived when the first photosynthesizing cells evolved— the *prokaryotes we know as blue-green bacteria (cyanobacteria). These single-celled, nucleus-free life forms

The color of these Yorkshire-Landrace pigs is due to hemoglobin molecules to which oxygen molecules have attached by linking to the iron atoms they contain. "White" humans are pink for the same reason.

application of this property is to the measurement of oxygen concentrations in artificial atmospheres, such as in incubators for premature babies: The magnetism of the gas is monitored and then interpreted in terms of the concentration of oxygen molecules.

Oxygen is highly reactive. When its molecules are torn apart (perhaps by intense heat), the liberated atoms can form strong bonds with atoms of other elements. The strength of these bonds is due to the smallness of the oxygen atom: The central nucleus can exert a strong force on neighboring electrons, including those of other atoms, because it can approach them very closely. Because it is highly reactive, oxygen exists in the atmosphere as *diatomic molecules,* which consist of two atoms linked together.

Huge quantities of oxygen gas are used in steelmaking, where about 1 ton of oxygen is needed to prepare 1 ton of metal. It is blown through the impure molten iron, combines with the impurities present (particularly carbon), and carries most of them away as gas. Oxygen is better for this purpose than air, which is mainly nitrogen (3) and which carries away too much heat when it is blown through the molten metal.

acquired hydrogen from water (discarding the oxygen) and carbon and oxygen from carbon dioxide, in the process of building their own carbohydrates (79). Thus, the oxygen that we now prize so highly, that is essential to most animal life, and that must be carried whenever we explore alien environments, was originally a pollutant in an atmosphere that favored a different form of life. That great pollution left its imprint on the earth, for the surge of oxygen that accompanied the emergence of photosynthesis *oxidized the iron dissolved in the seas. The earth rusted, and the great deposits of red iron ore record the epoch.

Oxygen itself is an odorless, colorless, tasteless gas that condenses to a pale blue liquid. The color change comes about when *pairs* of molecules cooperate in the absorption of light—a phenomenon that is possible only when the molecules are close together in a liquid. Oxygen also has the unusual property of being magnetic. This is most clearly shown by the ability of a magnet to pick up liquid oxygen, but the gas is also magnetic. One

Oxygen condenses at −183°C to a pale blue magnetic liquid. The color results from the interaction of neighboring molecules and is absent in the gas.

Nitrogen is the most abundant gas in the atmosphere, accounting for 78 percent of its volume, so that over three-fourths of every breath we inhale is nitrogen. This abundance of nitrogen probably outgassed from the rocks, just as did the other gases of the early atmosphere. However, nitrogen molecules are too heavy and too slow to have escaped the earth, and they are too unreactive to have combined with other substances to as great an extent as those of oxygen have; thus, what was abundant remained abundant.

Like oxygen, nitrogen forms diatomic molecules. However, the atoms in the N$_2$ molecule are bound together by a *triple* bond: N≡N. This results in one of the most strongly bonded molecules known, one that can survive collisions with other molecules which, for double-bonded oxygen, would have led to a reaction. Nitrogen's relative lack of reactivity allows it to act as a dilutant for the dangerous oxygen of the air. Without atmospheric nitrogen, one spark would long ago have ignited all the vegetation of the earth. Nitrogen's inertness must not, however, be confused with argon's. Argon's inertness can be attributed to its individual atoms having no tendency to combine with others. Nitrogen's inertness is a property of the *molecule*, not the atoms, and is due to its atoms having formed three strong bonds to each other. Once those bonds are broken, nitrogen is highly reactive and forms numerous compounds, as you will see.

Many of the molecules in living cells, particularly the *proteins, contain nitrogen atoms. Nitrogen is therefore essential to the growth of plants and crops. Its incorporation into life, first into plants and then into animals, begins with its conversion into nitrogen oxides (11, 12) by lightning flashing through the air and by solar radiation acting on the upper atmosphere. These more reactive compounds are then washed into the soil by rain.

A major highway for the movement of nitrogen from the atmosphere to the soil is *nitrogen fixation,* the incorporation of atmospheric nitrogen into compounds, which may be natural or contrived by industry. Biological nitrogen fixation is caused only by certain prokaryotes, including bacteria, particularly cyanobacteria (blue-green algae), and actinomycetes (branching, multicellular moldlike organisms). Some of these bacteria (particularly *Azotobacter* and *Clostridium*) can exist and operate individually; but the most important (*Rhizobium*) form symbiotic associations with higher plants, particularly the legumes (clover, pulses such as peas and beans, alfalfa, acacia), whose roots they colonize. In all cases the agent responsible for the fixation is the enzyme system *nitrogenase,* which consists of two protein molecules. One of these molecules is based on 2 molybdenum atoms, 32 iron atoms, and between 25 and 30 sulfur atoms (as well as an extensive scaffolding of carbon atoms). The other protein is based on iron. (It is sobering to think that this crucial enzyme system, which stands high in the food chain, requires the properties of molybdenum, an element that occurs as only a few atoms in every million.) Industrial chemists are intensely interested in the mode of action of nitrogenase, for if it could be mimicked, then the world would have a ready means for harvesting the nitrogen of the skies and redistributing it as fertilizer— or, more elegantly, for genetically engineering nitrogenase into a crop that would generate its own fertilizer as it grew.

Rhizobium leguminosarum nodules on the root of the pea, *Pisum sativum* (magnified 1.5 times).

Cyanobacteria, which obtain their energy from sunlight, their carbon from carbon dioxide, and their nitrogen directly from the atmosphere, are believed to be the first microorganisms that colonized land. This hypothesis is supported by the observation that they were the first to reestablish themselves after an eruption of the volcano on Krakatoa in 1883 had completely destroyed all life in an extensive surrounding area. Indeed some pioneering algae, specifically the red cyanobacteria that occur in huge colonies in the Red Sea, have their settlements named after them.

Biological nitrogen fixation is too slow and too local to support the burden of production we currently impose on the land, and additional nitrogen must be reaped from the air in huge amounts and applied as fertilizer. This is done by prying nitrogen atoms apart and allowing them to combine with hydrogen to form ammonia (7), which can be absorbed by the soil or converted into other fertilizers. Nitrogen lost from the atmosphere, naturally and otherwise, is replenished by the decomposition of vegetation and flesh, for when protein molecules decompose, the nitrogen atoms are released as ammonia molecules, which in due course degrade into molecular gaseous nitrogen.

. . . .

CARBON DIOXIDE   (4) CO_2

Carbon dioxide is the gas we exhale, for it is one end product of the consumption of the organic compounds we ingest as food. When an organic compound burns (and here I include that extremely sophisticated slow burning in living cells we call "*metabolism*"), each carbon atom is excised from its molecule by two oxygen atoms and is carried away as carbon dioxide. (If insufficient oxygen is provided, the carbon is carried away as carbon monoxide, CO.) In a flame, the disruption of the molecule and the formation of strong carbon-oxygen bonds is accompanied by the release of energy as heat. Carbon dioxide is the end of the road for the combination of carbon with oxygen and its formation corresponds to the maximum release of energy: Carbon dioxide is a dead form of carbon. But it is not inert, for green vegetation uses the energy of sunlight to pluck carbon dioxide from the skies, combine it with hydrogen that is obtained from water, and build its carbohydrates (79) in the process known as *photosynthesis*. This process and the dissolving of carbon dioxide in the oceans maintain the balance of carbon dioxide in the atmosphere (or so we hope).

In a muscle or a brain, the energy released when carbon dioxide is formed may be used to raise a weight or produce an idea. Carbon dioxide is also the end product of the partial consumption of carbohydrates during fermentation, an incomplete form of respiration that forms alcohol (27) as another principal product. Hence, carbon dioxide is the gas in the head of beer and the bubbles in champagne, for it comes out of solution when the bottle is opened and the pressure is released. Carbon dioxide in water is common as soda water or seltzer, and with added flavors, as beverages of various kinds. In water it forms the very weak acid *carbonic acid,* which tingles the tongue, is a taste enhancer, and acts as a mild bactericide. Carbonic acid is also said to encourage flow from the stomach to the intestine, which perhaps accounts for the rapid inebriating effect of champagne.

Carbon dioxide is the fourth most abundant component of the dry atmosphere and the most abundant gas in the atmospheres of Mars and Venus. A great deal of carbon dioxide was removed from the early atmosphere as the oceans fell from the skies, for the gas was dissolved in them. Now most of the carbon dioxide of the early planet lies beneath our feet, in the form of carbonate rock—chalk and limestone. No similar precipitation of water occurred on the hot surface of Venus, and that on Mars (if there was any) was insufficient, so on those planets the carbon dioxide remains in the atmosphere. It has been calculated that the mass of carbonate rock on earth, plus the amount of carbon dioxide in the atmosphere and dissolved in the oceans, is approximately the same as the mass of carbon dioxide that now hangs in the skies of Venus. Had the earth been only 10 million kilometers closer to the sun than its present 140 million kilometers, the temperature of its surface would have been too high for the oceans to form, and earth would have evolved into a planet like Venus.

Carbon dioxide in the atmosphere acts partly to trap the *infrared radiation emitted from the warm surface of the earth. Because carbon dioxide is transparent to the sun's visible light, that light can penetrate through the atmosphere to the earth's surface. As the surface warms,

Much of the original carbon dioxide of the planet has been trapped as carbonate rock, the compressed remains of shellfish. The shells are colored by the impurities, particularly iron ions, that they have incorporated.

it emits infrared radiation that cannot escape back into space because carbon dioxide molecules absorb it. This trapped energy warms the atmosphere in a process known as the *greenhouse effect*. In a real greenhouse the heat buildup is due more to the glass preventing a convective mixing of warm inside air with cold outside air than to the absorption of infrared radiation. This somewhat different effect in a real greenhouse was confirmed only recently, after the term "greenhouse effect" was already in use by climatologists and astronomers.

Carbon dioxide is used as a leavening agent in baking. Typical *baking powders* consist of sodium bicarbonate ($NaHCO_3$), an acid (or, typically, two acids, such as tartaric acid and the acidic salt, sodium aluminum sulfate), and starch, which acts as a filler and helps to separate the acid and bicarbonate particles and keeps them from reacting prematurely. But even so mundane a product as baking powder has an engaging technology, because it has to provide two separate bursts of action—that is, of carbon dioxide release. The first occurs at room temperature as a result of the action of the moistened tartaric acid, and it produces many tiny cavities in the batter. The second burst of activity is due to the action of the aluminum salt, and it occurs at high temperature. This second flux of carbon dioxide swells the cavities to give the desirable final light texture.

The carbon dioxide used in breadmaking is usually formed by the action of yeast on sugar or other small carbohydrate molecules. Such yeast particles are present in the air, but to achieve more uniform characteristics in baking, a particular strain, *Saccharomyces cerevisiae,* is normally cultured in dilute molasses and then used.

. . . .

OZONE   (5) O_3

Ozone is present in the upper atmosphere in the *ozone layer,* a band about 20 kilometers thick centered between 25 and 35 kilometers above the surface of the earth. If all of it were collected and compressed to the atmospheric pressure characteristic of the earth's surface, it would form a layer about 3 millimeters thick. The ozone is formed when the sun's *ultraviolet radiation is absorbed by molecules containing oxygen: Oxygen atoms are driven out of those molecules and subsequently bond to O_2 molecules that they strike. Once formed, the ozone molecule absorbs more ultraviolet radiation at a different wavelength and is blasted apart. Both processes, ozone formation and ozone decomposition, absorb radiation and hence help to protect the living organisms on the surface below. The absorption of ultraviolet radiation by the gas is so efficient that at wavelengths near 250 nanometers, in the ultraviolet, only 1 part in 10^{30} of the incident solar radiation penetrates the ozone layer. A being with eyes able to see only in 250-nanometer light would see the sky as being pitch black at noon.

Ozone is a blue pungent gas (*ozein* is the Greek for "to smell") that condenses to an inky blue-black explosive liquid. Its smell can be detected near electrical equipment and after lightning, since it is also formed by an electric discharge through oxygen. Because ozone would be encountered in airplane cabins on commercial flights at altitudes of about 15 kilometers and would cause coughing and chest pains, the incoming air is passed through filters that *catalytically decompose ozone to ordinary oxygen. Atmospheric ozone attacks the carbon-carbon double bonds in rubber (63) and contributes to its weathering.

WATER AND AMMONIA

Some important molecules consist of a single central atom surrounded by enough hydrogen atoms to satisfy its tendency to form bonds. Each of these molecules, which include the two described here and methane (16), can be regarded as the simplest member of a series of increasingly complex compounds in which the hydrogen atoms are replaced by other atoms or by groups of atoms. In water the central atom is oxygen, and in ammonia it is nitrogen. In each case the central element is temporarily protected from reaction by a shell of hydrogen atoms.

. . . .

WATER   (6) H_2O

Water occurs in huge abundance on the earth, where most of it lies in the great oceans that cover 71 percent of its surface to an average depth of 6 kilometers. This water dropped from clouds formed when the heat in the interior of the young earth drove oxygen and hydrogen atoms out of chemical combination in rocks built of compounds like *mica,* a form of potassium aluminum silicate. The newly formed molecules were brought to the surface in streams of lava and then released as water vapor, to form great clouds that could rain once the earth had cooled. Hence our oceans were once our rocks.

The oddest property of water is that it is a liquid at room temperature. This is surprising because so small a molecule would be expected to be a gas, like ammonia (7), methane (16), and its even closer relative hydrogen sulfide (H_2S, 101). Water's liquidity stems from the joint presence of the tiny hydrogen atoms, the lone pairs of electrons on the oxygen atom, and the power of the oxygen atom to attract electrons strongly. These features cause water molecules to be strapped together by networks of hydrogen bonds (page 11). One molecule can link with four others, and each of those four neighbors can link to others as well. As a result, the molecules cluster together as a mobile liquid, rather than move independently as a gas.

Water occurs as a liquid under normal conditions on earth, and it freezes to a solid at an unexpectedly high temperature because the links between the molecules are so strong. All three forms of water—ice, liquid, and vapor—are abundant on earth, but very little is in a form suitable for human consumption; 97 percent of it is too salty, and 75 percent of the earth's fresh water is solidified at the poles. The remaining 1 percent of the total water is drinkable, but most of that is inaccessibly deep groundwater. Thus, only 0.05 percent of the total, the water that runs through lakes and streams, is readily available. Some ancient groundwater is mined in deep wells, but that water recedes further from the surface as we use it but do not replace it.

Water's oddness does not end with its liquidity: Most solids are more dense than the liquids from which they freeze, but ice at 0°C is *less* dense than water at 0°C. As a result, ice floats on water, giving us icebergs and a solid skin on frozen ponds. This skin insulates the liquid water beneath, protects it from the cooling winds above, and can keep it from freezing during the winter. Thus, aquatic life can survive in the liquid, even though its

Icebergs (left) float on water because ice is less dense than water. Benzenebergs sink in benzene because solid benzene, like most solids, is denser than its liquid.

A model of the structure of ice.

roof is frozen. This quirk of density is again due to hydrogen bonds, for when water freezes, its molecules are held apart, as well as held together, by the hydrogen bonds between them: Each molecule grips its neighbors firmly, but at arm's length. The structure of the solid is therefore more open than that of the liquid, in which many of the hydrogen bonds have collapsed, so that it is less dense. Floating icebergs are thus a sign of the strength of hydrogen bonds; indeed, it was hydrogen bonds that brought death to the *Titanic.*

Water is also an excellent solvent. It mixes readily with alcohol (27) as in wine, beer, and spirits, because it can form hydrogen bonds with alcohol molecules. Sugar (81) dissolves readily in water for similar reasons. Many ionic solids dissolve in water, as salt and other minerals dissolve to yield the brine of the seas. This occurs because the water molecules simulate the surroundings of the ions in the crystal.

Water is a perfect medium for such processes as the transport of nutrients into cells and the carving of landscapes by flushing minerals out of rocks. Water appears to be essential for life, since it can provide a fluid environment within cells through which other molecules can migrate. Water can transport molecules up to cells, give them mobility within cells, and transport molecules away from cells to other locations (this includes the expulsion of molecules to the outside environment as waste). It can transport organic molecules like glucose (79) and the ions of such elements as sodium, potassium, and calcium that are so essential to an organism's functioning. Moreover, water, when it is a liquid, can do all this at body temperature. Fortunately, it cannot dissolve the calcium phosphate of our bones, so that our skeletons do not dissolve in our own fluids and we do not wilt.

The color of pure water—it is pale blue when a thickness of about 2 meters is viewed against a white background—is also due to its hydrogen bonds. When one molecule vibrates, it drags and pushes against neighbors to which it is hydrogen-bonded; as a result it absorbs a little red light, leaving a bluish hue in the light that remains. The same hint of blue is often seen in ice formations.

Opposite page: Water and ice in bulk is pale blue, as in this ice cave in Glacier Bay, Alaska. The color is due to the hydrogen bonding of the water molecules.

SIMPLE SUBSTANCES

AMMONIA (7) NH₃

The pungent gas ammonia takes its name from the Ammonians—worshippers of the Egyptian god Amun—who, like the fainthearted of later years, used *sal volatile* (ammonium carbonate, $(NH_4)_2CO_3$) in their rites. Ammonium chloride occurs naturally in crevices near volcanoes; it smells of ammonia because it vaporizes when it is warmed and decomposes into ammonia and hydrogen chloride.

Ammonia is one of the most important industrial chemicals, for it begins the chain of industrial food production. More molecules of ammonia are manufactured each year than of any other industrial chemical. Almost all these new ammonia molecules are made by the *Haber process,* in which atmospheric nitrogen (3) and molecular hydrogen (H_2) are forced to combine under high pressure (about 100 times atmospheric pressure) and temperature (about 700°C) in the presence of a *catalyst. This process for harvesting the air was invented by Fritz Haber during World War I when Germany was cut off from its normal supply of nitrates from Chile.

Although the conversion of atmospheric nitrogen to ammonia is called nitrogen *fixation* (3), it might more accurately be called nitrogen *mobilization*. In the process, the tightly held nitrogen atoms of N_2 are ripped apart, and each one is surrounded with less tightly held hydrogen atoms. The nitrogen is thereby made more susceptible to attack—attack that may result in its incorporation into amino acids (73–75) and thence into the proteins (76–78) of organisms. Hence, ammonia is manufactured in such mountainous abundance because it is a source of usable nitrogen. Its major use is in fertilizers, either as ammonia or after conversion first to nitric acid (14) and

then to nitrates. Ammonia is also the starting point for the incorporation of nitrogen atoms into industrially produced molecules, including nylon (72) and pharmaceuticals.

Ammonia is a colorless, flammable gas under normal conditions. It dissolves very readily in water because it can form hydrogen bonds with the water molecules. This high solubility contributes to the perceived odor of the gas, because ammonia is able to dissolve readily in the aqueous mucus that coats the olfactory epithelium of the nose (page 124). Water would probably smell just as pungent if our sensors were not constantly saturated by it. An overripe Camembert or Brie also smells of ammonia, for the gas is formed as the nitrogen-containing protein molecules (76) decompose. The same process takes place in stale urine (130).

Ammonia condenses at −33°C to a colorless liquid which resembles water in its ability to act as a solvent. Liquid ammonia has the interesting ability to dissolve some metals (including sodium, potassium, calcium, and, slightly, magnesium), giving blue solutions. The color is due to electrons trapped in cavities between the ammonia molecules. These fluids are solutions of electrons—almost potted electricity—and pouring them from one vessel to another is equivalent to pouring electrons.

Ammonia can be regarded as the parent of the compounds called *amines*. Amines often retain the pungency of ammonia but frequently in a modified and more disagreeable form. To see the kind of change that occurs, you should turn to putrescine (131) and its partner cadaverine, as one day you certainly will.

SMOG, POLLUTION, AND ACID RAIN

We have described air as perhaps it ought to be, with innocuous, essential components. However, air as it truly exists is quite a different thing, for it contains alien molecules that stem from natural and "civilized" sources that use the sky as a sewer. Some of the more common pollutants that often inhabit city skies are described in this section. Although chemistry undeniably contributes to pollution (particularly in the hands of the economists, politicians, physicists, farmers, homeowners, and others who make use of the benefits it brings), the deleterious properties of the molecules described here (and those discussed in the section entitled "Nasty Compounds") should be weighed against the colossal contributions of the other molecules described in this book.

SULFUR DIOXIDE   (8)   SO$_2$

SULFUR TRIOXIDE   (9)   SO$_3$

(8)

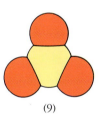

(9)

Just as carbon dioxide is formed when oxygen carries off carbon atoms from burning carbon and organic compounds, so sulfur dioxide is formed as a colorless, dense, toxic, nonflammable gas with a suffocating odor when the yellow solid element sulfur burns in air. The same end product is obtained from the combustion of compounds containing sulfur atoms, including some of the components of oil and coal. Sulfur dioxide also belches out of the ground at volcanoes and emerges from plants where iron and copper ores (pyrite, FeS$_2$, and copper sulfide, CuS) are heated during the extraction of the metal. It is also formed where volatile sulfur compounds produced by the decay of plant and animal matter are *oxidized in the air. About 200 billion kilograms of sulfur reach the sky each year from industrial sources, join-

ing there the 300 billion kilograms that nature donates.

Sulfur dioxide is used as a preservative in a number of foods and beverages, for it is able to combine with the oxygen that would otherwise attack the commodity being preserved. The molecules responsible for the colors and flavors of dried fruits and fruit juices survive longer if sulfur dioxide is present. Sulfur dioxide is used in winemaking, partly to suppress the growth of wild yeasts and bacteria that would sour the grape juice into vinegar (32) and partly to prevent *oxidation. More is used with white wine than with red so that oxidation of the pale yellow pigment is prevented.

Huge numbers of sulfur dioxide molecules are manufactured deliberately, by burning sulfur, in the production of sulfuric acid (10). They are then encouraged by a

Elemental sulfur consists of crown-shaped rings of eight sulfur atoms. Depending upon how the rings stack together, they form blocklike *rhombic sulfur* or needlelike *monoclinic sulfur*. The former is the stable form at room temperature.

*catalyst to combine with another oxygen atom so that most of the manufactured sulfur dioxide is converted into sulfur trioxide. That conversion takes place only slowly without a catalyst, as when atmospheric sulfur dioxide forms the trioxide in water droplets. The oxygen-rich trioxide reacts vigorously with many substances and is usually converted to sulfuric acid as soon as it appears.

. . . .

SULFURIC ACID   (10) H_2SO_4

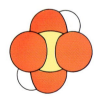

Sulfuric acid (which when pure is a viscous, oily liquid) is the world's most widely used industrial chemical, and no other chemical is manufactured in greater tonnage. (More *molecules* of ammonia are produced, but each one is lighter.) Almost every manufactured item comes into contact with the acid at some stage. So important is sulfuric acid that its annual production has been taken as a measure of the degree of industrialization of a country and of commercial activity in the world. However, there has recently been a shift in its use: Most of it is now going into the production of phosphate fertilizers, and it is increasingly being replaced by hydrochloric acid in the surface treatment of iron. Hence, it is now probably more useful as an indicator of agricultural—rather than industrial—activity.

Sulfuric acid also falls from the skies, for it is one of the acids of *acid rain*. It is formed there when sulfur dioxide (8) is converted to sulfur trioxide (9) in airborne droplets of water.

These nudibranchs (sea slugs) secrete sulfuric acid for defense.

NITRIC OXIDE   (11)   NO

NITROGEN DIOXIDE   (12)   NO₂

(11)

(12)

Both these oxides are formed by the combination of nitrogen and oxygen in the combustion chambers and hot exhausts of automobile and airplane engines, and they are collectively known as NO_x. Nitric oxide is formed first, and it and its reaction product, nitrogen dioxide, build up before dawn and in heavy traffic. At dawn, when the sun's radiation begins to stimulate chemical reaction, ozone (5) is produced.

Nitrogen dioxide is a deep orange-red gas that, together with smokelike particles, is responsible for the color of smog. The color of nitrogen dioxide and many of the colors that adorn the world more spectacularly are due to the process of *absorption;* they must be distinguished from colors due to emission (page 17) and *in-

candescence. How colors arise by absorption is explained on page 149, and that explanation applies to the color of nitrogen dioxide as well as to the colors of vegetation.

Nitric oxide molecules are manufactured on a large scale, since they are precursors of nitric acid (13). The starting material is ammonia (7), which is allowed to react with oxygen in the presence of a *catalyst. This is actually a controlled form of ammonia burning in which some oxygen atoms carry away the ammonia's hydrogen atoms as water molecules while one oxygen atom remains, captured by the nitrogen atom. If the resulting nitric oxide is cooled and mixed with more oxygen, it extracts another oxygen atom from an O_2 molecule and forms the brown dioxide.

NITRIC ACID   (13)   HNO₃

Nitric acid is formed in the atmosphere when nitrogen dioxide dissolves in water, and it falls from the skies as a component of acid rain. When pure, it is a colorless liquid; the straw color familiar from school laboratory

bottles of dilute acid is due to nitrogen dioxide formed by the acid's partial decomposition. It is manufactured by dissolving in water the nitrogen dioxide produced by the oxidation of ammonia.

The acid is the parent of the *nitrates,* which are ionic solids that contain the nitrate ion. Among these is ammonium nitrate (NH_4NO_3), which is widely used as a fertilizer because it contains a high proportion of nitrogen and (like almost all nitrates) is soluble in water. The solubility of nitrates accounts for their rarity as minerals, because they simply get washed away. However, *Chile saltpeter* (impure sodium nitrate), which is found in large deposits in the arid regions of Chile, is an exception. Its origin is uncertain, but it may be the result of weathering on animal or vegetable remains. *Guano,* a phosphate deposit that also contains nitrates, was once used as a fertilizer; it is the excrement of fish-eating sea birds and is found in large quantities on the dry, rocky islands off the coast of Peru.

When the hydrogen atom of nitric acid is replaced by a carbon atom and the atoms the carbon atom carries, the properties of the molecule often change dramatically. To see what I mean, jump to TNT (154).

HYDROGEN PEROXIDE   (14) H_2O_2

The additional oxygen atom has little effect on the physical properties of hydrogen peroxide in comparison with water, but it profoundly changes the chemical properties. Hydrogen peroxide is a powerful oxidizing agent, destroying organic compounds that come into contact with it. That is partly its role in pollution, for fragments of the peroxide (such as HO_2 and HO) occur in photochemical smog, where they attack unburned fuel molecules and convert them into lachrymators such as PAN (15). This chemical activity is put to use as a bleach for hair and, on a larger scale, for paper pulp. The action of hydrogen peroxide as a bleach stems from its ability to oxidize and destroy pigments, including the melanin (142) responsible for the color of black, brown, and fair hair. Many such colors are due to molecules with strings of carbon atoms joined by alternating single and double bonds. The hydrogen peroxide molecule attacks the double bonds: After breaking one of the bonds open, it dumps its excess oxygen between the fragments, forming a three-membered ring

called an *epoxide.* With the alternating sequence of single and double bonds disrupted, the color disappears. An advantage of hydrogen peroxide over some other bleaching agents, such as chlorine gas, is that its decomposition products, oxygen and water, are not pollutants.

Hydrogen peroxide is present in trace amounts in honey, where it is the product of an *enzyme that oxidizes glucose (79). Its presence accounts for the antibiotic action that once led to honey being used for dressing wounds. There are, however, better and more hygienic sources of hydrogen peroxide than bees.

PEROXYACETYL NITRATE (15) C₂H₃O₅N

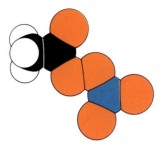

Peroxyacetyl nitrate (PAN) is a derivative of hydrogen peroxide in which one hydrogen atom has been replaced by a CH₃—CO— group and the other has been replaced by an —NO₂ group. With its —O—O— peroxide link (14) and other oxygen atoms, PAN can be expected to be an irksomely active molecule.

PAN occurs in photochemical smog. It is the powerful lachrymator that is responsible for much of the eye irritation caused by smog. It brings tears to the eyes because the tear ducts respond to invasion by secreting a saline fluid (which also includes various antibodies in case the invasion is bacterial) in an attempt to wash the invader away. PAN is produced when bright sunlight causes reactions among fragments of unburnt hydrocarbon fuel molecules (20), which provide the —CH₃ group; oxygen, which provides the —O—O— link; and nitrogen dioxide, which provides the —NO₂ group. PAN is also largely responsible for the damage that smog does to vegetation; its very high oxygen content causes the *oxidation (in effect, partial combustion) of any organic matter it touches.

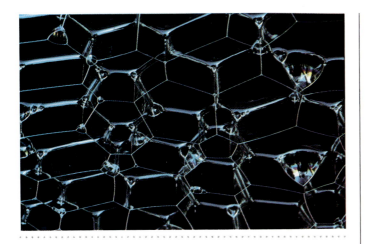

Soap bubbles between glass.

2

FUELS, FATS, AND SOAPS

Now we come to the principal types of molecules that will be the focus of our attention in the remainder of the book. With these organic molecules you can begin to see the web of intricacy that carbon spins, largely by linking to itself. The structural prototype of all organic compounds is methane. This potent molecule can be extended into chains, rings, and networks of linked carbon atoms on which atoms of other elements, most notably oxygen, can hang.

Many of the molecules that spring structurally from methane have an enormous impact on the world, being deployed by nature, modified and synthesized by industry, and used by everyone. Some are merely burned as fuels. Two common but important types of molecules—alcohols and carboxylic acids—are used by nature and by chemists to forge links between chains of carbon atoms. You will read about their combination—esters—in many places throughout the book. There are esters in the fats and oils that act as food reserves in our bodies, that we consume by the ton during a lifetime, and that we do not always fully discard. Some fats we spill, and for those you will see how chemists engineer molecules that can infiltrate grease yet dissolve in water and hence can act as soap.

NATURAL GAS AND LPG

The replacement of gas manufactured from coal by natural gas, the gas trapped in porous rocks, transformed the pattern of energy usage during the 1950s, and now natural gas is pumped through pipelines that cross countries and continents. Much of that gas is used for domestic and industrial heating, but some is also used as the starting point for the synthesis of organic compounds. This section describes some of the typical components of natural gas and of the liquefied petroleum gas (LPG) used for mobile applications, including camping. All the molecules discussed are *hydrocarbons*, compounds containing only carbon and hydrogen.

Natural gas, like petroleum oil, originated from decayed organic matter, probably as the result of bacteria (typically cyanobacteria) scavenging for oxygen atoms and leaving strings of hydrocarbon molecules as residue. For such bacterial action to dominate oxidation, there must be a high productivity of organic matter and a low concentration of competing oxygen. These conditions are satisfied in sedimentary layers in coastal waters, where marine organisms thrive. The hydrocarbons that are produced are eventually squeezed out of the compacted muddy sediments into neighboring porous rocks—most often sandstone (particles of quartz and silicates bound together by a cement of silicate or carbonate) or limestone (calcium carbonate). Loss of the hydrocarbons by means of upward migration is prevented if the rock formation is roofed over by an impermeable layer: A typical hydrocarbon trap consists of an *anticline* (a convex upward fold) in which sandstone is capped by an impermeable shale (a fine-grained sedimentary rock composed of compressed silt and clay). The gaseous hydrocarbons rise highest in the porous rock, lying above the petroleum that fills the remaining pores and floats on the ground water in the pores below.

....

METHANE   (16)   CH$_4$

A single central atom of an element surrounded by hydrogen atoms constitutes a primitive but portentous molecule. In no case is this better demonstrated than when the central atom is carbon, as in *methane* [for the etymology of the name, see methanol (26)]. Carbon's tendency to form four bonds is satisfied if four hydrogen atoms are attached, as in this tetrahedral molecule. However, as you will see, methane is the precursor of all organic chemicals in which these hydrogen atoms are successively replaced by other atoms and by groups of atoms. Primitive, simple methane is the most pregnant of molecules.

Methane is an odorless, nontoxic, flammable gas. It is a gas of carbon atoms in flight, but each carbon atom is prevented, by its casing of hydrogen atoms, from immediately reacting and from sticking together with others to form a solid block of carbon. Methane is formed naturally when bacteria release single carbon atoms from digested organic material, and it occurs as *marsh gas* when it is not trapped by a suitable rock formation. It is the main component of natural gas, accounting for about 95 percent of the gas from midcontinental sources, 75 percent of the gas from the Texas panhandle, and nearly 70 percent of the gas from Pennsylvania.

In a gas flame in which methane combines with oxygen, the hydrogen atoms are ripped off the tetrahedral methane molecules by the oxygen to form water molecules (6), and the remaining carbon atom picks up oxygen atoms to form carbon dioxide (4). The blue-and-green light of the flame is emitted by energetic C_2 and

CH molecules formed briefly in the course of the combustion. In a limited supply of air, the carbon may be oxidized incompletely, forming carbon monoxide (CO) and a smoke of unburned carbon particles in which billions of atoms have stuck together to give *soot*. Then the flame is yellow with the glow of light emitted by the *incandescent carbon particles.

The blue and green colors of the methane flame are due to the excitation of an electron to a new, higher-energy location in the C_2 and CH molecules. The electron falls back to its original location almost immediately and, in doing so, discards the excess energy as light. A similar process is involved when an electric discharge passes through argon (1) and neon, but the colors emitted are different because different amounts of energy are discarded. That blue and green light is emitted by C_2 and CH shows that they discard more energy than neon when the electron falls back.

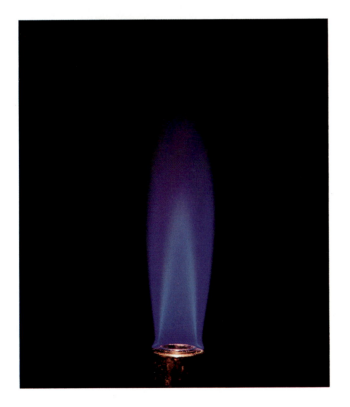

The colors of a gas flame are due to energetically excited atoms and molecules discarding their excess energy as light.

ETHANE (17) C_2H_6

Now we begin to allow the methane molecule to grow. A simple adaptation is to break one C—H bond and insert a —CH_2— group between the carbon and hydrogen atoms. This slightly more complicated hydrocarbon is *ethane,* and with it carbon begins to show its ability to combine with itself—an ability that can result in magically complex molecules.

Ethane, a colorless, odorless gas, accounts for about 30 percent of the natural gas from Pennsylvania and is also extracted from oil wells. Ethane molecules interact slightly more strongly with each other than methane molecules (this is a common result of increasing the number of atoms, and hence electrons, in a molecule); although it is a gas at room temperature, it condenses to a liquid at a higher temperature ($-89°C$) than methane ($-162°C$).

Much more important than ethane is its cousin ethylene (47), to which most of the ethane in natural gas that is not simply burned is converted.

PROPANE (18) C_3H_8
BUTANE (19) C_4H_{10}

(18)

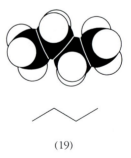

(19)

Both propane and butane are used as LPG and are carried around as camping gas. Both are gases at normal temperatures, but butane condenses to a liquid at 0°C and so cannot be used for camping under cold conditions. Thus, butane tends to be used in the southern United States, and propane (which condenses at $-42°C$ under normal pressure) in the north. Butane is also used as a fuel in cigarette lighters where, under pressure, it is stored as a liquid. For both, complete combustion is like that of methane, with the end products being carbon dioxide and water.

More than 20 percent of the propane obtained from natural-gas sources is converted to ethylene (47) and its relative propylene. Butane's destiny is often synthetic rubber (64).

GASOLINE AND COAL

A polarized light view of light heating oil. The colors arise from the interference of rays that take different times to pass through the liquid and show regions of different density and molecular packing.

The two dominant *fossil fuels* of the twentieth century are gasoline, derived from the liquid remains of decomposed marine organisms (as described on page 34), and coal, the compressed remains of decomposed vegetation. In each case the organic remains have survived without being *oxidized because they have been sheltered from atmospheric oxygen beneath deep rocks. Petroleum oils, from which gasoline is *distilled, and the products of the distillation of coal are the raw materials of the petrochemical industry. They are used in the manufacture of many of the molecules discussed later, since they already consist of chains and rings of carbon atoms that can be used as building blocks in chemical processes.

. . . .

OCTANE (20) C_8H_{18}

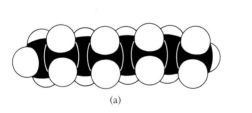

(a)

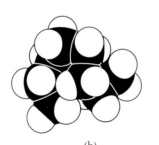

(b)

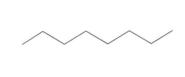

The octane molecule results when we continue the hypothetical process that led from methane (16) to ethane (17). Now a sufficient number of —CH_2— units have been introduced into the original C—H bond of methane to make the chain eight carbon atoms long (hence *oct*ane).

Hydrocarbon molecules that contain half a dozen or so carbon atoms interact just strongly enough with each other to give a liquid at room temperature, so they are convenient to transport in tanks. But the liquid is still volatile and not too viscous to form a fine spray in the carburetor of an engine. Octane is representative of the size of the hydrocarbon molecules present in a gallon of gasoline [see the discussion of octane number following the description of (21)]. Diesel fuel is less volatile: Its molecules are typically hydrocarbons with about sixteen carbon atoms in a chain. The sixteen-carbon analogue of octane is called either *cetane* or, more formally, *hexadecane.*

Octane is called a *straight-chain hydrocarbon* because all its carbon atoms lie in a line, with no branches. But do not be misled into thinking that "straight-chain" means geometrically straight. Every hydrocarbon chain is actually a zigzag of carbon atoms and is flexible as well;

moreover, each atom can be twisted around the bond joining it to its neighbor. A gallon of gasoline therefore contains some octane molecules that are rolled up into a tight ball (20b), others that are stretched out but still zigzag (20a), and others in the various intermediate configurations. Octane molecules are constantly writhing and twisting, rolling and unrolling, so that a gallon of gasoline is more like a can of molecular maggots than a box of short sticks.

. . . .

2,2,4-TRIMETHYLPENTANE (21) C_8H_{18}

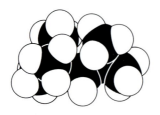

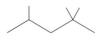

The name 2,2,4-trimethylpentane is a descriptive label indicative of the structure of this molecule. Because it is such a mouthful, the molecule is usually (and more affectionately) known as *isooctane,* the name we shall use from now on. (Like octane itself, it has eight carbon atoms.)

Isooctane is an example of a *branched-chain hydrocarbon,* for its carbon chain has "side groups" like the branch lines of a railroad. It has the same molecular formula as octane and hence the two are *isomers—molecules built from the same numbers of each kind of atom, but linked together differently.

Isooctane and other branched hydrocarbon molecules are more desirable than straight-chain molecules in commercial gasolines. A straight-chain hydrocarbon ignites explosively in an engine (perhaps because its chain of carbon atoms is so exposed), causing "knocking" and power loss. A branched-chain hydrocarbon, however, burns smoothly, heating the gas in the cylinder throughout the power stroke, and so exerts a steady pressure on the receding piston. The ability to burn smoothly is measured as the *octane number* of the fuel. Isooctane, with excellent nonknocking properties, is ascribed an octane number of 100. Heptane [which is like octane (20) but with only seven carbon atoms], knocks horribly and is

ascribed a value of 0. A 95-octane fuel is then equivalent in knocking characteristics to a mixture of 95 percent isooctane and 5 percent heptane by volume.

The corresponding scale for diesel fuel involves the *cetane number* [cetane is hexadecane, as mentioned in (20)]. A cetane number of 100 corresponds to pure cetane, and 15 (the lowest point on the cetane-number scale) corresponds to pure heptamethylnonane, one of cetane's branched-chain isomers in which seven CH_3—groups are strung out like warts along a nine carbon atom chain. It is worth noting that, for a diesel engine, a rapidly igniting *straight*-chain molecule is desired. This reflects the different operating principles of the two types of engine: In a diesel engine, the fuel is sprayed in during the power stroke (not, as in a gasoline engine, all in one shot prior to compression) and needs to ignite as it enters the cylinders.

Gasoline companies strive to increase the proportion of branched-chain molecules in their products through *catalytic reforming.* In this process, straight-chain molecules are heated in the presence of a *catalyst; this causes groups of atoms to break off the straight-chain hydrocarbons and then reattach elsewhere, forming branches on the molecules.

TETRAETHYLLEAD   (22) $Pb(C_2H_5)_4$

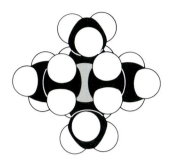

Tetraethyllead is an example of an *organometallic* compound, one in which one or more organic groups are linked to a metal atom. In this case the molecule can be pictured as consisting of four *ethyl* (CH_3—CH_2—) groups anchored to a lead atom in a tetrahedral arrangement. The molecule consists, in essence, of a single atom of lead buried deep inside a hydrocarbon environment.

The tetraethyllead molecule is useful not for the lead atom it carries, but as a ready supply of ethyl groups. The lead-carbon bonds are frail, and the four ethyl groups break off their anchor in the hot combustion chamber of an internal-combustion engine. They then take part in the combustion of the hydrocarbon fuel, promoting smooth burning through *radical chain reactions*. In such reactions, a *radical* (a molecule with an unpaired electron) attacks another molecule, thereby producing another radical that can go on to attack yet another molecule, and so on. The ethyl group is in fact such a radical, and it enters into the chain that results in combustion within the engine.

Unfortunately, lead poisons both people and the catalytic converters fitted to their cars. The latter are there to complete the combustion of partially burned fuel to carbon dioxide and water before it escapes to the atmosphere and is converted to pollutants by peroxide (14). Lead additives are therefore doubly banished, on the one hand being declared illegal by governments and on the other falling necessarily into disuse.

BENZENE   (23) C_6H_6

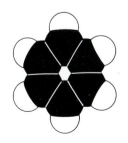

The planar, hexagonal arrangement of carbon and hydrogen atoms characteristic of benzene plays a very special role in chemistry because it is exceptionally resistant to attack. It appears in many of the larger organic molecules, acting as a kind of relatively inert plinth to which other groups are attached. The reasons for its stability are

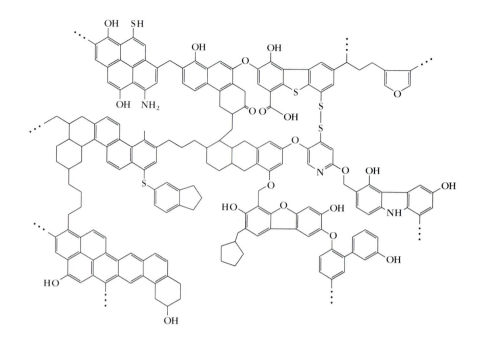

A schematic version of a portion of the molecular structure of coal. Note the large number of benzenelike rings. These sheets break up when the coal is heated.

complex, but they include the lowering of energy that results when bonding electrons are able to spread around the ring of carbon atoms.

Benzene was once called *benzol,* but this name is now reserved for the less pure grades of benzene. It is relevant to the present discussion because the combustion characteristics of hydrocarbon fuels are improved if compounds based on benzene are included in gasoline. These compounds are collectively known as *aromatic compounds,* for many, including benzene, have characteristic (but not always pleasant) odors. They are produced by catalytic reforming of petroleum (page 38) and, to a much smaller extent currently, by the distillation of coal.

Coal is an extremely complex mixture of compounds with an internal structure like that depicted in the illustration above. Note the many benzenelike rings; when coal is heated in the absence of air, these great sheets

vibrate and, ultimately, break up. The molecular fragments can be driven off, collected, and separated. The proportions of the different types of molecules in the product depend on the temperature to which the coal has been heated; the smallest fragments, including carbon monoxide and methane, are obtained at the highest temperatures, when the violence of the breakup is greatest.

The residue left after this *distillation* process is called *coke.* It is formed by heating coal to 350–500°C, at which point it softens and partly decomposes, and then raising the temperature to about 1000°C. Coke is mainly carbon but includes some mineral and remaining volatile matter. A great deal is used in blast furnaces in the manufacture of iron from its ores; different coals are blended to obtain a coke of sufficient strength to withstand the weight of ore in the furnace.

TOLUENE (24) C_7H_8

XYLENE (25) C_8H_{10}

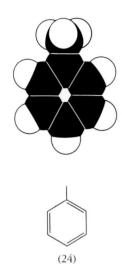

(24)

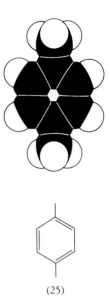

(25)

The toluene molecule can be viewed as a benzene mole-cule in which one C—H bond has been broken open to accept a —CH$_2$— group. The xylene molecule is then a toluene molecule in which that process has been repeated elsewhere in the benzene ring. In fact there are three different xylenes, with the second —CH$_2$— group in-serted either next to the one already present, or one car-bon atom away from it, or diametrically opposite it across the ring. Only this last *isomer is shown.

Both these aromatic hydrocarbons occur in gasoline, and together with benzene they are the main contribu-tors to high-performance *BTX* gasoline. Their concen-trations are enhanced by catalytic reforming in an oil refinery. Toluene, which is also the basis of TNT (154), is so called because it was originally obtained from *Tolu balsam*, the yellow-brown, pleasant-smelling gum of the South American tree *Toluifera balsamum*. This balsam has been used in cough syrups and perfumes.

All three xylenes are liquids with characteristic odors. They were once obtained by distilling wood in the ab-sence of air (*xulon* is Greek for "wood"). When not being burned in engines, para-xylene (the isomer shown) is used for the production of artificial fibers (66). A mix-ture of the three isomeric xylenes is called *xylol* and is used as a solvent.

ALCOHOL AND VINEGAR

"Alcohol" is a general term denoting an organic compound that contains the group —OH. The origin of the name, from the Arabic *al Kohl* for "the fine powder," is rather curious. The Egyptians stained their eyelids with the black inorganic compound antimony sulfide. This was obtained by grinding, but eventually the name was applied to the essence of anything, including the liquid obtained by distilling wine. This section describes a series of compounds related to wine and its *oxidized product vinegar, and introduces three more types of compounds that figure frequently in the following pages: *aldehydes,* which are compounds that contain the group

carboxylic acids, which are organic compounds that contain the *carboxyl* group

and *esters,* which are combinations of alcohols and carboxylic acids.

....

METHANOL   (26)   CH$_4$O

Methanol, or methyl alcohol, is also known as *"wood alcohol,"* because it was originally obtained by distilling hardwoods. (*Methy* is a Greek word that usually means "intoxicate" but sometimes means "wine," and *hule* is the usual Greek word for "substance" but is sometimes used for "timber" or "group of trees".) Methanol also occurs in wood smoke, and traces are present in new wine, where it contributes to the bouquet. Although methanol does cause an initial inebriation when it is drunk, it is an indirectly poisonous liquid. Its toxicity is largely due to the formation of formic acid (31) and formaldehyde (29) by the enzyme *alcohol dehydrogenase* in the body. These compounds attack the *ganglion cells in the retina, cause degeneration of the optic nerve, and can cause permanent blindness. Death usually follows ingestion of about 50 milliliters or more.

Methanol is a colorless, mild-smelling, almost tasteless liquid. It is highly flammable and burns with an almost invisible blue flame because there is insufficient carbon in each molecule to aggregate into incandescent soot particles (page 35).

ETHANOL (27) C_2H_6O

GAMMA-AMINOBUTANOIC ACID (28) $C_4H_9O_2N$

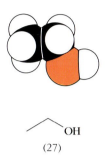

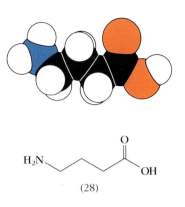

(27)

(28)

Ethanol, or ethyl alcohol, is what the world calls *"alcohol."* Since antiquity (though neither universally nor without remission), it has been the essential component of most socially acceptable intoxicating liquids. Physiologically, however, it acts as a depressant, like a general anesthetic. Ethanol seems to its imbiber to be a stimulant, but in fact it acts by freeing parts of the *cortex from inhibitory controls.

To understand the action of ethanol, it is necessary to delve into the brain and examine the role of the neurotransmitters, the molecules that communicate between nerve cells at synapses (nerve-cell junctions). Several of these neurotransmitters will be mentioned in later sections, but here we focus on gamma-aminobutanoic acid (GABA).

The GABA molecule normally acts by binding to a protein molecule on the surface of the "upstream" nerve cell of a synapse, and *inhibits* the cell's activity. This inhibition results from a distortion the GABA causes in the local structure of the cell membrane, in essence wid-

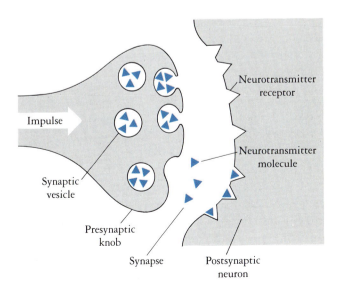

A synapse showing the presynaptic vessels containing the neurotransmitter and the postsynaptic binding sites to which they migrate when they are released.

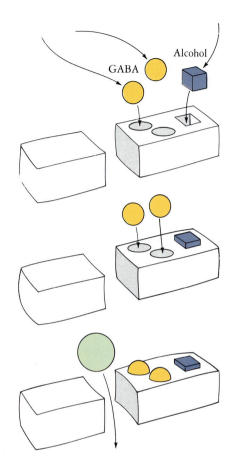

The white blocks represent protein molecules that surround channels through which chloride ions may pass into the cell. Each protein carries sites to which alcohol (ethanol) and GABA molecules may bind and modify the shape of the protein and hence the size of the channel.

ening the channels through which chloride ions (Cl⁻) can pass into the cell. The higher internal concentration of these ions changes the voltage difference across the membrane, and the cell becomes unable to fire. The ethanol molecule binds to the same protein as a GABA mol-ecule, but at a neighboring location, and distorts it slightly with the effect that the GABA can bind to it more easily. Hence, the presence of ethanol in the fluid surrounding the synapse indirectly influences firing and the voltage across the nerve cell membrane by enhancing the binding of GABA and thereby opening up the chloride ion channels. Some tranquilizers and sedatives act in a similar way; for example, the benzodiazepines [such as Valium, (148)] link to the same protein molecule as ethanol and GABA, but at a different location. Because the benzodiazepines and ethanol attach to the same protein molecule, there is a strong synergistic interaction between them: Valium can be lethal when taken with alcohol.

Other physiological effects that accompany the ingestion of ethanol include its interference with the production of some antidiuretic hormones, which leads to enhanced water secretion, urination, and hence a sense of dehydration. Alcohol also causes blood vessels to dilate; the resulting increased flow of blood through the capillaries beneath the skin gives a rosy hue to the complexion, along with a feeling of warmth.

Ethanol is made by fermentation of carbohydrates (79) and, industrially, from ethylene. It mixes with water in all proportions, since its —OH group can form *hydrogen bonds with the water molecules. Wine, beer, and spirits all contain ethanol, and for table wines about one molecule in twenty is ethanol (the bulk of the remainder being water). *Proof* in the United States denotes twice the percentage of ethanol by volume, so that 100 proof spirits are 50 percent ethanol. Spirits contain numerous other components as well; Scotch whisky, for example, includes volatile organic molecules from the smoke of *peat*, which is decaying organic matter that is burned in the process of drying the barley malt (partially germinated kernels) before it is fermented. *Vodka* (Russian for "little water") is virtually only ethanol and water: It can be prepared from the cheapest sources, usually grain, because most of the flavor is removed by passing it through charcoal. Ethanol metabolism in the liver produces acetaldehyde (30) and, as explained on page 46, hangovers.

FORMALDEHYDE (29) CH$_2$O

The formaldehyde molecule can be thought of as being obtained by removing two hydrogen atoms from a methanol molecule—one from the carbon atom and one from the oxygen atom. That is, in fact, the origin of the name *aldehyde* (from *al*cohol *dehyd*rogenated) given to compounds like formaldehyde that contain the group

(normally denoted —CHO). It also describes the industrial preparation of formaldehyde, in which oxygen is used to pluck two hydrogen atoms off the methanol molecule and carry them away as water. Billions of kilograms of formaldehyde are produced in this way each year, for its high reactivity makes it a valuable *industrial intermediate,* a substance that can readily be converted to a wide variety of other substances.

Formaldehyde (more formally, methanal) is a pungent, colorless gas with a suffocating odor. It is soluble in water and is often encountered as the 40 percent aqueous solution called *formalin,* which is used for preserving biological specimens. Formalin retains its sterilizing capacity even when it is diluted to 10 percent, and it is used to kill anthrax spores during the treatment of wool and hides. Formaldehyde is also present in wood smoke and is one of the agents responsible, by attacking bacteria, for the preservative action of the smoking of foods. The mode of action in all these cases appears to be its ability to react immediately with the —NH— and NH$_2$— groups characteristic of proteins (76) and to link together neighboring protein chains, hardening the substance and taking the protein molecules out of commission. This same type of reaction is used to produce synthetic resins and adhesives [as with urea (130), to obtain *urea-formaldehyde resins*}. The preserving and sterilizing action of formaldehyde can therefore be thought of as the formation of resins from living starting materials.

Methanol is metabolized to formaldehyde by the *enzyme *catalase.* Because that enzyme is also involved in the chemistry of vision (page 146), it occurs in the retina of the eye, and a significant amount of metabolism occurs there. The formaldehyde that is produced crosslinks the retinal proteins, removes them from active participation in vision, inhibits oxygenation of the retina, and for doses in excess of about 2 grams results in blindness.

Smoking coats food, including the salmon and trout shown here, with various bactericides, including formaldehyde and phenols of various kinds. Smoked fish are often artificially colored to enhance their appeal; kippers are dyed with "Kipper brown," a mixture of synthetic dyes.

ACETALDEHYDE (30) C₂H₄O

Acetaldehyde, a pungent, colorless liquid that boils at about room temperature, is the primary metabolic product of ethanol on its route to becoming acetic acid (32). It is produced by the enzyme *alcohol dehydrogenase*, which occurs primarily in the liver but also to a small extent in the retina. Generally, the bigger the person, the bigger the liver and the quicker the alcohol is metabolized and removed from circulation. Acetaldehyde is one of the chemical agents responsible for a hangover, although there are also numerous complicated and interrelated contributions from the physiological changes that occur as the body responds to unnaturally high ethanol levels and the mild narcosis, acid imbalance, and dehydration it induces. Alcohol dehydrogenase is present in our bodies because we need to metabolize the alcohol produced in small amounts by the normal digestion and breakdown of carbohydrates, and in large amounts by the bacteria in our intestines.

Some substances (notably other alcohols) compete with ethanol for alcohol dehydrogenase. These alcohols are present in *fusel oil* (from the German word for 'rotgut" or, more specifically, "evil spirit"), a byproduct of fermentation, and hence are present in some distilled spirits. The metabolism of wine and spirits is therefore slower than that of vodka, which is largely free of everything except ethanol and water.

Acetaldehyde is also a product of the action of *Saccharomyces cerevisiae,* a yeast which is allowed to develop on fino and amontillado sherries and imparts to them their nutty flavor. It also contributes to the odor of ripe fruits.

Cells of the yeast *Saccharomyces cerevisiae.* This "brewer's yeast" is used in breadmaking (page 101).

FORMIC ACID (31) CH$_2$O$_2$

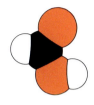

The "formic" in formic acid's name reflects its origin: Formic acid is one component of the venom injected by stinging ants (*formica* is Latin for "ant"). It is also a component of the fluids injected by stinging caterpillars. (The sting of the common nettle, *Urtica*, like that of the hornet, is largely due to a mixture of histamine and acetylcholine: Note the convergent evolution of weaponry.)

It is also a product of the metabolism of methanol by means of formaldehyde, analogous to the conversion of ethanol to acetic acid (32) by acetaldehyde. The heightened acid levels due to this formic acid production damage proteins and contribute to the internal devastation that methanol causes.

This ant (*Polyrhachis,* from the subfamily *Formicinae*), which is seen here with larvae and cocoons, injects a venom rich in formic acid. The nettle (*Urtica*), like the hornet and the octopus, adopts a different strategy and uses a mixture containing acetylcholine and histamine.

ACETIC ACID   (32) $C_2H_4O_2$

Acetic acid, a colorless liquid with a sharp odor, is the acid component of *vinegar* (from the French words *vin aigre,* meaning "sour wine") and is responsible for its characteristic smell. Since acetic acid is an *oxidized form of ethanol, it is produced when wine stands exposed to air and the ethanol undergoes aerobic oxidation by the bacteria *Acetobacter.* It is also what gives poor wines their vinegary taste. Acetic acid is also produced in dough leavened with the yeast *Saccharomyces exiguus,* which cannot metabolize the sugar maltose; as a result *Lactobacillus sanfrancisco,* a particular group of bacteria that depends on maltose, can thrive and excrete acetic acid and lactic acid (33). The dough becomes *sourdough* and is baked into the bread for which the San Francisco area is known.

LACTIC ACID   (33) $C_3H_6O_3$

Lactic acid is a solid at room temperature. Of itself, that fact is not very interesting; however, it does serve to illustrate how a second —OH group provides new opportunities for *hydrogen bonding between neighbors, and hence a stronger attraction between them. Slightly more interesting is the existence of two kinds of lactic acid molecules, an L form (for *levo,* from the Latin word for "left") and a D form (for *dextro,* from the Latin word for "right"). Each is the mirror image of the other; and just as a left glove cannot be superimposed on its mirror image, a right glove, so L-lactic acid cannot be superimposed on D-lactic acid. The two forms have virtually identical chemical properties but differ in their effect on plane polarized light: L-lactic acid rotates its plane of polarization—the plane in which the electric field lies—to the right, and D-lactic acid rotates it to the left as the light approaches the observer. This phenomenon is known as *optical activity* and is a general property of *chiral molecules*—molecules that cannot be superimposed on their mirror images. As you will see, chirality is far from being of only academic interest (74, 117, 158).

D-lactic acid is obtained by the action of bacteria on meat extract, and L-lactic acid from the fermentation of sucrose with *Bacillus acidi levolactii*. A mixture of the two is obtained from sour milk.

Lactic acid is widespread in nature. A clue to its ubiquity is obtained by comparing its molecular formula, $C_3H_6O_3$, with that of glucose (79), $C_6H_{12}O_6$: A lactic acid molecule is, in essence, half a glucose molecule. Indeed, a widespread source of lactic acid is the anaerobic fermentation of sugars and the action of *enzymes on glucose supplies. Fresh milk (page 57) rapidly becomes populated with bacteria that act on the milk sugar *lactose*, break it up for its energy, and excrete lactic acid. The acid causes the fatty droplets to coalesce, and the milk curdles. This process is encouraged in a controlled way in the production of yogurt, which depends on lactic acid production by a mixed culture of the bacteria *Lactobacillus bulgaricus* and *Streptococcus thermophilus*. Fermented pickles also owe their tartness to lactic acid. *Sauerkraut* is produced by steeping fresh cabbage in brine, which suppresses the growth of some bacteria and gives others—first *Leuconostoc mesenteroides* and then *Lactobacillus plantarum*—a chance to thrive. As they do, they consume glucose units and excrete lactic acid, sharpening the taste of the cabbage.

Lactic acid is also produced from glucose by enzymes in our sweat glands (which accounts for the acid taste of perspiration) and by the action of bacteria on the lining of the vagina, which is a rich store of glucose. It is also produced, as a last resort, in muscle that has exhausted its immediate oxygen supply and cannot metabolize glucose aerobically. Thus a sprinter may unconsciously resort to the anaerobic energy supply of his or her ancestors, and make do with the energy released by slicing glucose molecules into lactic acid halves. Unfortunately, this builds up the acid concentration in the muscles, interfering with their operation so that they feel heavy and weak, and may produce cramp.

Bodily lactic acid becomes involved indirectly with heavy drinking: Since the metabolizing capacity of the liver may be saturated by the demands of the ethanol, lactic acid is not removed so efficiently. It may then build up in the bloodstream, raise the acidity of the muscles, and lead to the kind of fatigue experienced—for different reasons—by an athlete. Lactic acid can also influence the deposition of solid compounds (specifically, salts of uric acid, a derivative of the class of molecules known as *purines*). These are normally excreted in the urine, but their excretion is inhibited by lactic acid and they may, instead, deposit in the joints causing the painful condition known as *gout*. The first deposits usually occur in small joints, particularly the metatarsal-pharyngeal joint (of the big toe); they are encouraged by alcohol and foods that are rich in purines, including the classic accompaniments of cartoon gout—claret and port.

Left- and right-handed molecules of lactic acid exist and crystallize into colonies of the same type. The presence of the two types is shown by different colors when a sample is examined with a polarizing microscope.

FATS AND OILS

There is no deep, fundamental distinction between a fat and a fatty oil: At room temperature a fat is a solid and an oil is a liquid, and what is a fat in one house may be an oil in another. The alcohol part of most fats and oils is *glycerol* (the commercial product is also known as *glycerin*). The carboxylic acid part generally has a long chain of carbon atoms and is called a *fatty acid*. Different fats and oils correspond to different fatty acids being attached to the glycerol unit, and each combination has its own melting point. The fats and oils from different sources actually consist of different mixtures of these molecules.

GLYCEROL (34) $C_3H_8O_3$

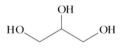

A glance at the structure of glycerol suggests that it will be markedly different from its parent, the gas propane. In particular, the three oxygen atoms sensitize their hydrogen atoms for *hydrogen-bond formation, and the glycerol molecules stick together in a viscous, syrupy liquid. Moreover, not only do the molecules bond strongly to each other, but they also form strong hydrogen bonds with any water in the vicinity. What is less obvious is that the hydrogen bonds and the oxygen atoms give the molecule a sweet taste; but that property is best dealt with later (page 106).

Glycerol's ability to bond water molecules to itself accounts for its widespread use in cosmetics, pharmaceutical preparations, and foods. It is used as an *emollient* (a softener) and a *demulcent* (a soother) in cosmetics, and as an antidrying medium in toothpaste. It is added to candies to prevent their crystallization and is sprayed on tobacco to act as a humectant, to keep the leaves moist and prevent them from crumbling before processing. Glycerol is also added to glues to prevent them from drying too fast, and it is incorporated in plastics—notably cellophane (85)—as a *plasticizer*, to keep them supple by acting as a lubricant between polymer molecules.

The sweetness and smoothness (essentially, viscosity) of the liquid and its solutions makes glycerol a desirable component of wines. Nature contributes to our pleasure in this respect through the unlikely agency of "noble rot," caused by the fungus *Botrytis cenerea*, which under certain conditions attacks grapes on the vine, injures

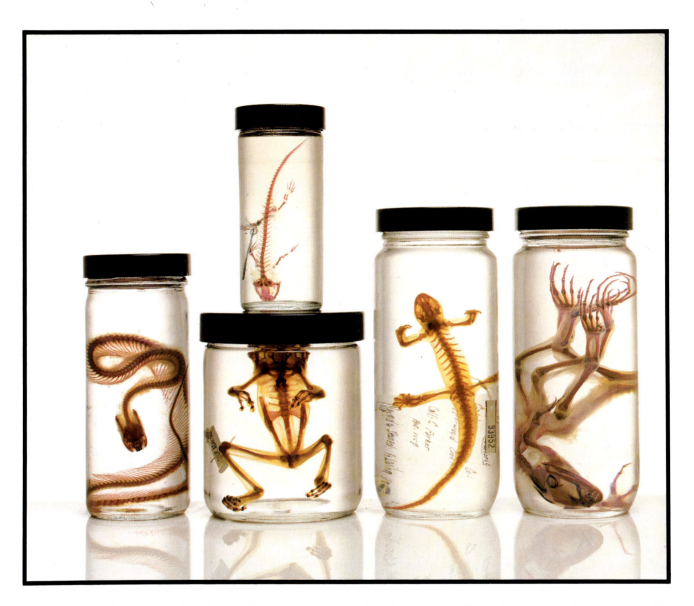

A selection of vertebrates preserved in glycerol. Glycerol penetrates the cells and thus preserves the specimens without drying them out. The refractive index of glycerol also makes the tissues more visible. These specimens have been chemically transformed to render the flesh clear, the bone red, and the cartilage blue.

Zinfandel grapes attacked by *Botrytis cenerea* (noble rot).

their skins, and allows some water to evaporate. The botrytized grapes are left richer in glycerol, and the resulting wines (most notably some Sauternes) are sweet and smooth to the palate. Glycerol is also present in red wines, particularly good Burgundies. The formation of "legs" on the side of a glass after the wine has been swirled, which is a sign of quality, is due to the presence of slow-draining, viscosity-enhancing glycerol.

Apart from its ability to anchor water molecules, each —OH group of glycerol can combine chemically with and anchor one carboxylic acid molecule. The resulting esters are the fats and oils described in this section. If only one —OH group is esterified, the ester is a *monoglyceride;* if two, a *diglyceride*; and if all three, a *triglyceride*.

. . . .

STEARIC ACID (35) $C_{18}H_{36}O_2$

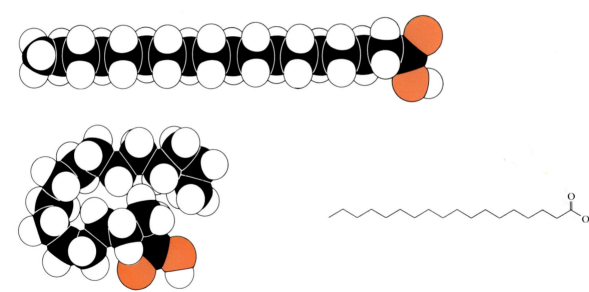

Stearic acid is an example of a fatty acid. In particular, it is a typical *saturated* carboxylic acid. In a saturated compound there are no double bonds between neighboring carbon atoms. One consequence of stearic acid's having only single carbon-carbon bonds is that its hydrocarbon chain is flexible [like that of octane (20)], because each —CH₂—CH₂— unit acts as a hinge. Hence, as well as stretching out into a zigzag, stearic acid can roll up into a compact ball.

TRISTEARIN (36) C$_{57}$H$_{110}$O$_6$

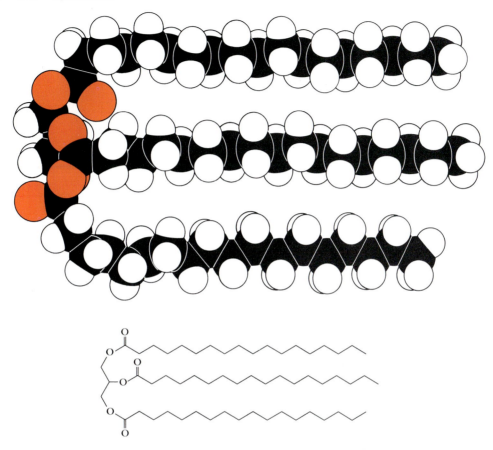

The tristearin molecule is one of the most complicated we have met so far, but its structural theme is very simple. The molecule is a triglyceride (34); think of it as a threefold *ester in which each of three stearic acid molecules is anchored to a glycerol —OH group. The glycerol chain can be seen on the left of the illustration, with the three hydrocarbon chains of the fatty acid spreading away to the right like three streamers. In a normal sample, the streamers would be coiled and tangled with each other and with the streamers of neighboring molecules.

Some of the properties of tristearin can be guessed by glancing at its molecular structure. One important feature is the absence of hydrogen atoms attached to oxygen

atoms, leading us to expect that the molecules do not form hydrogen bonds. This is so, and it is the reason why fat and water do not mix. That is an advantage for organisms, for then fat molecules are less likely to be washed out of their bodies and can therefore be stored. Moreover, since their molecules do not stick together tightly, we can also expect fats to be less dense than water, and to float on it. Nevertheless, tristearin molecules can maximize their packing by rolling their flexible fatty acid side chains up into moderately compact balls; hence tristearin is a fat at body temperature, not an oil.

Compounds that are predominantly hydrocarbonlike often dissolve in hydrocarbonlike solvents such as benzene (23). Naturally occurring substances that dissolve in hydrocarbons and alcohols but not in water are called *lipids*; fats and oils are lipids.

Tristearin is one of the principal components of beef fat and cocoa butter (the major component of chocolate). The role of fat is principally as a reserve fuel supply; when used it is ultimately *oxidized to carbon dioxide and water. It is therefore interesting to note that animals use a fuel very similar to that adopted for automobiles: Apart from their slightly longer carbon chain, the side chains of fats differ from diesel fuel only in that they are already very slightly oxidized (as a result of the presence of the six oxygen atoms in tristearin).

The length and saturation of the stearic acid chains is an advantage to animals, since fat need not be stored in a special tank and can be used as an insulating coat. The camel's hump is an elegant solution to the double problem of storing energy and water, for as noted above, tristearin oxidation produces water when it is oxidized.

Fats have various secondary roles in food. They act as solvents for many flavor components and for some of the molecules that cause color. Beef fat, for example, is colored slightly yellow by the carotene molecule (136) acquired from grass. Fats also increase the satiety value of meals, since they leave the stomach slowly and hence

Cacao is the source of chocolate and cocoa. The pods grow along the branches and on the trunk of the tree and contain the seeds known as "beans." These "beans" were once used by the Aztecs as currency.

delay the onset of hunger. (This is also why drinking milk before alcohol slows the latter's absorption; it takes longer for the alcohol to reach the intestine.) And just as the automobile designer uses long-chain hydrocarbon molecules to lubricate moving parts, so animals use fats to lubricate their own meat fibers. A part of the tenderness of beef is due to lubrication by tristearin and its analogues.

. . . .

OLEIC ACID   (37) $C_{18}H_{34}O_2$

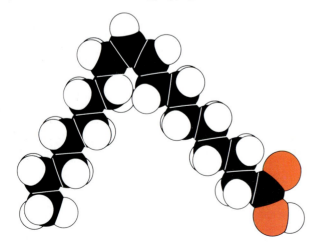

Oleic acid is an *unsaturated* fatty acid, since it has a double bond between two neighboring carbon atoms. In general, an unsaturated compound is one with at least one carbon-carbon multiple bond; it is so called because it does not have its full complement of hydrogen atoms.

The double bond has a marked effect on the shape of the molecule and of the triglycerides it forms with glycerol. Since the molecule cannot be twisted around the double bond, the chain is much less flexible than that of stearic acid and cannot roll up into a ball. The molecules of the esters it forms are much less compact than tristearin, the molecules do not pack together so well, and the compounds are oils rather than fats. Salad oils are often "winterized" (chilled and filtered) to remove the fats that harden and lead to cloudiness at refrigerator temperatures. Pork, lamb, and poultry fats contain a higher proportion of unsaturated fats than beef and hence are softer to the touch.

Plant stems and leaves are under less evolutionary disadvantage from weight than animals are and can store their food reserves as carbohydrate. However, their seeds need to support the developing embryo until it is self-sufficient, and for this a compact, efficient food reserve is required. Hence, oils are more abundant in seeds than in stems. Corn, cottonseed (*Gossypium*), soybean (*Glycine max*), safflower (*Carthamus tinctoris*, the petals of which also supply the red dye used for some kinds of rouge), and the solar-tracking sunflower (*Helianthus annuus*) are all good sources of unsaturated fatty acids, and oleic acid itself is the principal fatty acid in the olive oil pressed from the ripe fruit of the olive (*Olea europaea*).

The oleic acid present in the cocoa butter of chocolate gives it an interesting property. Most fats and oils are complex mixtures of triglycerides, with the result that they do not melt at a precise temperature but rather soften over a range. Cocoa butter, however, is markedly uniform in composition: Each triglyceride molecule has an oleic acid streamer springing from the central carbon atom of the glycerol anchor, and the two other streamers are often either stearic acid or the closely related palmitic acid. This uniformity results in a much sharper melting point than is common for fats, and chocolate is brittle almost up to its sharp melting point of 34°C (just below

body temperature, 37°). Moreover, the melting is so sudden and energy-absorbing (like any melting) that when it occurs in the mouth it gives a feeling of coolness. One problem with cocoa fat is that the molecules can stack together in six ways, giving solids with different melting points. The conversion of the solids to the low-melting variety occurs when chocolate is subjected to temperature variation. This gives rise to a grayish "bloom" on the otherwise hard, shiny surface.

Linolenic acid, which occurs in linseed oil (produced from flaxseed, *Linum usitissimum*), is closely related to oleic acid but has three double bonds and so is an example of a *polyunsaturated* fatty acid. The stiffness of the molecule that results from these multiple bonds leads us to expect the triglycerides built from it to be oils. We would also expect the oil to be chemically reactive, since carbon-carbon double bonds are chemically sensitive regions of molecules and are liable to attack by other substances.

Linseed oil is in fact an example of a *drying oil*, one that produces a film when its double bonds are attacked by oxygen on exposure to air. The mechanism involves first the creation of *radicals by oxygen attack, followed by a *chain reaction that results in links being formed between neighboring molecules. The process, which is called *polymerization*, occurs in polyunsaturated vegetable oils that are left to stand and is accelerated when they are used in frying. Old frying oil should not be topped off with new oil, for surviving radicals in the former will cause polymerization to proceed in the latter.

Air-induced polymerization is desirable for oil-based paints, which consist of a pigment in suspension in a drying oil, traditionally linseed oil plus some catalyst (drier) to accelerate the oxidation. The setting qualities of alkyd paints (69), which include polyunsaturated fatty acid components, depend on the development of cross-linking between chains of those acids.

The fatty acids that occur in fish oils such as cod liver oil are rich in polyunsaturated fatty acids. This is probably the outcome of evolutionary pressure, since the triglycerides they form pack so poorly together that they remain liquid even in the cold environments the fish inhabit.

CHOLESTEROL   (38) $C_{27}H_{46}O$

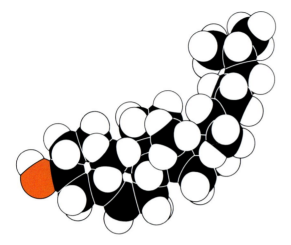

The ingestion of saturated fats provides the body with long, flexible chains of carbon atoms. It can use them to produce cholesterol, and hence raise the concentration of cholesterol in blood plasma to above its usual level. Note that although this molecule has an elaborate, rigid, hydrocarbon framework, its business end (more formally, its *functional group*) is primarily the —OH group. In other words, cholesterol is chemically an elaborate alcohol (hence the *-ol* in its name).

Cholesterol, which is normally produced in the liver, plays an essential role in metabolism in the body. It is a precursor of various hormones (159), is used in cell-membrane production, and is the starting point for the production of bile compounds (its name comes from the Greek words for "bile solid"). These compounds act as follows. When lipids enter the duodenum, they are partially broken down into fatty acids by the *enzyme pancreatic lipase*. But the long-chain fatty acids characteristic of fats and oils are too much like hydrocarbons to be soluble in water; left to themselves, they would precipitate as a greasy solid and clog the intestine like a blocked drain. The bile compounds, however, act as detergents (page 61), surrounding the fatty acid droplets and scattering them as groups of about a million molecules,

giving a clear dispersion that can flow through the gut and can be absorbed and stored as fuel.

So much for the good and biochemically necessary side of cholesterol. Most people think of the shady role of cholesterol as the encourager of *atherosclerosis* (from the Greek words *atheroma*, meaning "porridge," and *skleros*, meaning "hard"), a porridgy deposit on the intima, or smooth walls of arteries. These deposits are an accumulation of *lipids, mainly cholesterol, and complex carbohydrates; they are hardened as calcium ions accumulate in them from the blood plasma seeping through. As the porridge hardens, it blocks the flow of oxygenated blood to the myocardium and can lead to ischemic (oxygen deficiency) heart disease.

There is some evidence that unsaturated fats are less likely to enhance the cholesterol level than saturated fats; apparently this involves a change in the way cholesterol is distributed between the plasma and various cellular compartments where it is stored and not a change in its rate of formation or decay. Manufacturers of edible fats are faced with the problem of maintaining a desirable level of unsaturation without producing a runny, inconvenient oil that could rapidly become rancid.

BUTTER AND MARGARINE

Butter is prepared from milk by churning its cream, which causes the fat globules to coalesce; the liquid that remains is called *buttermilk*. The original margarine (from the Greek word *margaron*, for "pearl") was invented as a butter substitute in 1869 and was a mixture of milk, chopped cow's udder, and beef fat. Today's margarine has somewhat more obscure origins, being synthesized from substances that include vegetable oils and petrochemicals (chemicals obtained from petroleum). Some of the components of butter and margarine, including the decomposition products that contribute to rancidity, are discussed in this section. The molecule responsible for the color of butter and margarine is carotene (136).

. . . .

BUTANOIC ACID (39) $C_4H_8O_2$

The fats and oils in cow's milk are rich in triglycerides built from short-chain fatty acids, including butanoic acid. Because the variety of short chains can be pushed past each other reasonably easily, butter is soft rather than waxy like fats [such as tristearin (36)] and the long-chain hydrocarbons called *paraffin waxes*. However, milk is a complex mixture that includes stearic acid (35) and oleic acid (37) fats. As in other animal fats, the complexity of the mixture results in a solid that melts over a range of temperatures rather than remaining hard up to a sharp melting point.

Human milk fat is rich in linoleic acid (41) and contains an as-yet-unidentified substance that promotes the growth of the bacterium *Lactobacillus bifidus*. The bacteria produce large amounts of lactic acid (33) in the infant's intestine, and the acidity suppresses the growth of harmful bacteria.

In milk, the fats form an emulsion with water, in which they are dispersed in small droplets. Each droplet is surrounded by detergentlike molecules (44), including cholesterol (38); these have a hydrocarbon part that can mingle with the hydrocarbon chains of the fats and one or more —OH groups that can form *hydrogen bonds with the surrounding water. They stabilize the fat droplets by surrounding them with a water-favoring sheath. Coagulation of the droplets occurs because of a protein present in the milk, which links the droplets together and, as they coalesce, they form a head of cream. Heating milk also causes a protein, lactalbumin, to coagulate; a skin is formed as water evaporates. The skin is moderately impervious to steam, and enough pressure can build up beneath it to cause the milk to overflow from the vessel. Some of the protein is destroyed and creaming is reduced by heating the milk briefly to about 100°C.

Cream on the top of milk is less common now that *homogenized milk* is widely available. Homogenized milk is formed by forcing milk through small openings under pressure. This breaks the fat globules into smaller particles, resulting in a more viscous, whiter, blander tasting, and more stable emulsion.

BUTANEDIONE (40) $C_4H_6O_2$

Molecules that contain the *carbonyl group* >C=O are called *ketones* and are responsible for many natural flavors and odors (page 124). Butanedione (also known as *diacetyl*) is a volatile yellow liquid ketone with a cheeselike smell. It is, in fact, the molecule that gives butter its characteristic flavor and the molecule you should have in mind when you smell it; for when cream is incubated with bacteria, they produce some butanedione. After incubation, the cream is churned. This breaks down the sheaths around the fat droplets, and they coalesce into a soft, solid mass. Sheep's milk and goat's milk are richer in short-chain triglycerides than cow's milk, and cheese made from them (such as Roquefort) is richer in pungent molecules.

You may be able to smell butanedione by sniffing your armpits or someone's unwashed feet, because it is a contributor to the odor of fermenting perspiration. Fresh sweat is almost odorless, but the action of the bacterium *Streptococcus albus,* which is present on the skin, increases its acidity and makes it an inviting feast for other bacteria; they, in turn, excrete pungent compounds including butanedione. Deodorants act by killing the bacteria.

LINOLEIC ACID (41) $C_{18}H_{32}O_2$

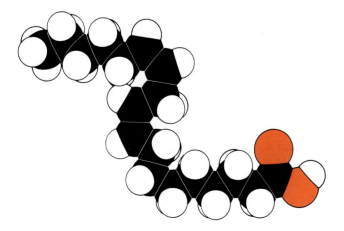

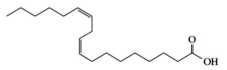

FUELS, FATS, AND SOAPS

Linoleic acid is the principal fatty acid in many vegetable oils, including cottonseed oil, soybean oil, and corn oil. It is also abundant in rapeseed oil (from members of the mustard family, *Brassicaceae*) and is used in the manufacture of margarine, shortening, and salad and cooking oils.

Triglycerides (36) built from linoleic acid are oils because of the double bonds in their chains of carbon atoms. Since margarine and shortening manufacturers want a soft solid, they bubble hydrogen through the oil in the presence of a nickel *catalyst. This *hydrogenation* process brings about several changes, including the partial saturation of the carbon chains as hydrogen atoms attach to carbon atoms that were originally joined by double bonds. The replacement of carbon-carbon double bonds with single bonds allows the carbon chains to become flexible. As a result, the molecules can pack together closely and the oil is converted to a fat. The hy-drogenation process is stopped sooner if the oils are destined to become softer (tub) margarines.

The elimination of the double bonds during hydrogenation also reduces the likelihood of attack by oxygen, so that the fat remains fresh longer. Nasty-smelling molecules are removed by passing superheated steam through the molten fat. This also removes the molecules responsible for color, so carotenes (136) of various kinds are added to restore a butterlike appearance. The odor of butter is simulated by adding butanedione (40). The flavor is enhanced and sharpened by *emulsifying the fats with skimmed milk that has been cultured with bacteria that produce lactic acid (33). The nutritional value is improved by the addition of vitamins A and D. And, finally, natural *surfactant molecules (*lecithins,* which are triglyceridelike substances with one side chain containing a phosphatelike group) are added to ensure that the entire concoction hangs together.

The seeds of the rape plant (*Brassica napus*) provide rapeseed oil, a source of linoleic acid.

from a load of wash is highly nutritious and can promote the growth of microorganisms in rivers and lakes. This can lead to *eutrophication,* or overnourishment, which leads to clogging by organic growth, perhaps to the point of transforming a lake into a swamp. *Brighteners* are fluorescent dyes that absorb some ultraviolet light and reemit it as visible light, hence making fabrics look brighter.

Proteolytic (protein-molecule destroying) enzymes excreted by *Bacillus subtilis* and *B. licheniformis,* which can survive the conditions encountered during laundering, are also included in some detergents. Their action is sufficiently specific so that they will modify the composition of the proteins in dirt particles without affecting the fibers of the fabric.

. . . .

SODIUM STEARATE   (43) $C_{18}H_{35}O_2Na$

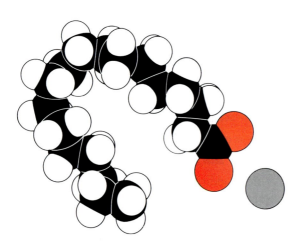

Tallow, the mixture of fat obtained by treating cattle fat with steam and collecting the material that floats on the water, contains tristearin (36). When tristearin is heated with sodium hydroxide (caustic soda), the stearic acid chains are broken off the glycerol (34), and the resulting mixture contains sodium stearate. In a detergent or soap, the long hydrocarbon chain is the part of the stearate ion that mixes with grease, and the ionic $—CO_2^-$ group is the head group that bonds with water.

Other fats and oils are also used in soaps—especially coconut oil, which contains a high proportion of lauric acid, a fatty acid with eleven carbon atoms in its tail. A typical toilet soap consists largely of fatty acid *salts made from tallow and coconut oil and so contains a high proportion of sodium stearate and sodium laurate. The combination of a potassium soap with excess stearic acid gives a slow-drying lather and is used in shaving soap.

SODIUM ALKYLBENZENESULFONATE (44) C$_{18}$H$_{29}$SO$_3$Na

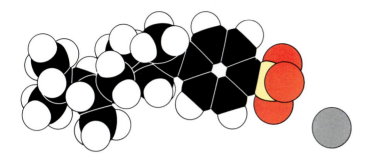

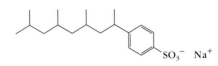

Synthetic detergent molecules like this one are an improvement on soaps for the following reason: When soap (43), a carboxylic acid derivative, is used in *hard water* (water containing calcium and magnesium ions picked up as it trickled through limestone hills), insoluble calcium and magnesium carboxylate *salts precipitate out and form an unpleasant *scum*. This does not happen with sulfonate detergents because calcium and magnesium sulfonates are more soluble in water than the carboxylate salts.

SODIUM PARA-DODECYLBENZENE SULFONATE (45) C$_{18}$H$_{29}$SO$_3$Na

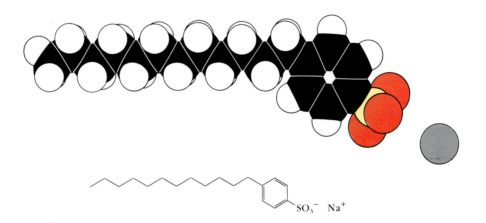

The dodecylbenzene sulfonate ion shown here is very similar to the ion shown in (44), the only difference being that the twelve carbon atoms of the side chain are strung out linearly and not branched. However, that difference has important ecological consequences: One problem with early detergents like the one in (44) is that when they run as effluent into rivers, bacteria are unable to destroy their branched side chains and the rivers foam. This has been overcome by synthesizing detergents with unbranched hydrocarbon chains, for such molecules can be digested by bacteria under aerobic conditions. That is, they are *biodegradable*. They also have just as good detergent powers as their branched isomers.

POLYOXYETHYLENE (46) $C_{14}H_{30}O_2$

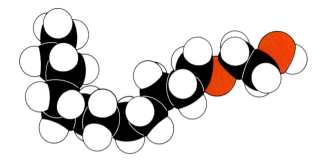

The head group in this molecule consists of the —O—CH$_2$—CH$_2$—OH group, which links to water by means of the *hydrogen bonds it can form: At least two oxygen atoms are needed to ensure that the molecule interacts strongly enough with water, despite its long hydrocarbon chain. The advantage of this *nonionic detergent* is that it is less efficient at stabilizing (or maintaining) foam than the ionic varieties (44, 45) and so results in less foam when it is used. It is also more effective at removing grease and soil at low temperatures. The former property makes it suitable for use in washing machines, and the latter leads to the economical use of energy, since cooler water can be used. The —O—CH$_2$—CH$_2$— group is sometimes repeated several times in the same molecule to enhance the water-attracting power of the molecule, and the hydrocarbon chain can take various forms.

Leopard coat.

3

SYNTHETIC AND NATURAL POLYMERS

Much of nature's artistry depends on the ability to spin complexity from simplicity by linking small, mobile, and easily transportable molecules into chains and webs. The resulting linked molecules, or *polymers,* are fibers, sheets, and blocks that we know as rubber, silk, hair, and wood. Chemists have sought to understand and emulate nature in this as in so many other ways and have achieved, in some substances, a passable imitation. In others they have exceeded nature in designing molecules for their special purposes. Now, not only do polymers sprout from skin as hair and wool and exude from insects as silk but they are also carried in truckloads from factories as plastics, textiles, and coatings.

Nature deploys two types of building blocks with awesome virtuosity: It spins life itself from amino acids and foods and structural materials from carbohydrates that are, in effect, plucked from the sky by photosynthesis. Chemists have tinkered with molecules in attempts to understand and improve on nature, and some of the fruits of their work are shown on the following pages. You will see how chains are formed and how the compositions and shapes of the chains determine the properties of the polymers. You will see why some plastics form strong fibers, why some are fully transparent but others merely translucent, and why some are tough and strong whereas others are soft. Now you will begin to touch and wear the everyday with a new understanding.

POLYMERS AND PLASTICS

Take away synthetic polymers and you take away a high proportion of the artifacts of everyday life. Here are a few of them.

The synthetic substances that have transformed the world during the twentieth century are the synthetic plastics. These substances have in common that they are built by stringing many small molecules together. The individual small molecules are called *monomers* (from the Greek words for "one part"), and the chains and nets they form are called *polymers* (for "many parts"). In some cases, two or more different types of monomer are linked together; the resulting material is then called a *copolymer*.

Polymeric materials are so important and commonplace that we shall devote this and the following five sections to them. Some are synthetic; others were happened upon by nature and then developed to a high level of sophistication.

ETHYLENE (47) C_2H_4

POLYETHYLENE (48) $(CH_2CH_2)_n$

(47)

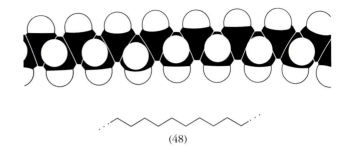

(48)

The ethylene molecule is the parent of an exceptionally important class of compounds, and we shall use it as a seed just as we used methane (16) to develop the hydrocarbons described so far. This molecule can be regarded as an ethane molecule (17) in which one hydrogen atom has been removed from each of the two carbon atoms, which then form a double bond.

The presence of the double bond in ethylene (which is also more formally called *ethene*) makes it much more interesting than ethane. A minor but important use for ethylene is in the ripening of fruit, for plants generate the gas when they are ready to ripen. It appears to stimulate the metabolic processes involved, perhaps by dissolving in and increasing the permeability of cell membranes. Fruit shippers often transport their produce unripe and then expose it to ethylene gas at its destination.

Ethylene is much more reactive than ethane, since the carbon-carbon double bond can open when attacked by other substances and form bonds with them, leaving the carbon atoms joined by a single bond. When hydrogen links with the molecule in this way, ethane is formed, which is not very interesting; the real importance of the double bond is in the molecules other than hydrogen that can add to it to give many different compounds. Because of its chemically tender double bond, ethylene is too reactive to be found in appreciable amounts in natural gas. It is formed during the refining of crude oil,

especially when the bigger hydrocarbon molecules are torn apart during the process known as *cracking*.

One of the substances that can add to ethylene is ethylene itself. *Polyethylene* is formed when ethylene molecules combine, and the process can continue until the string of linked —CH_2—CH_2— units has grown to an enormous length—perhaps thousands of units long. When you touch a polyethylene article, you can feel the characteristic waxy texture of a hydrocarbon.

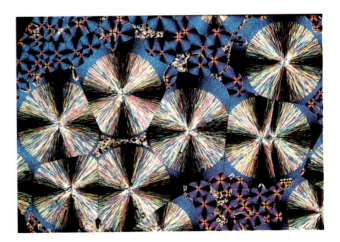

A thin film of polyethylene photographed using polarized light. The pattern arises from the presence of *spherulites,* or regions where the polymer molecules have aggregated into spheres.

In a typical sample of polyethylene there are molecules of many different lengths; each chain has many side chains—some containing a thousand carbon atoms—where the polymerization process has led to attack on an existing polymer chain. All the molecules are tangled together into a microscopic version of a plate of hairy spaghetti.

Pure polyethylene is translucent for the same reason a slurry of ice is. In the latter, the numerous small ice crystals lie at random orientations to each other, and light passing through is scattered in many different directions. This effect can be so powerful that some substances look brilliantly white even though they are composed of a colorless, transparent substance: Milk is one example, and white house paint, which contains colorless, transparent titanium dioxide, is another. In polyethylene, domains in which the molecules lie next to each other in an orderly fashion (so-called *crystalline regions*) alternate with domains in which the chains are jumbled together in a random way (*amorphous regions*). The crystalline regions lie at random orientations to each other, and they scatter the light like the crystals in ice.

For *high-density polyethylenes*, the reaction conditions are chosen so that the carbon chains are 10,000 to 100,000 carbon atoms long, are of reasonably similar lengths, and have fewer than one side chain per hundred carbon atoms. The molecules then pack together more effectively, so that the solid is denser, more crystalline, and stiffer than ordinary polyethylene.

Polyethylene has excellent electrical insulating properties. These partly stem from the tightness with which the electrons are trapped in their C—C and C—H bonds, so that a current cannot flow through the solid. They also reflect the inability of water and ions to penetrate into the oil-like hydrocarbon interior of the solid. Moreover, the molecules are uniformly electrically neutral, with no regions of enhanced positive and negative charge, unlike nylon (72), for example. As a result, they barely respond to electric fields. In particular, they do not begin to oscillate when exposed to an alternating electric field and so do not absorb and dissipate its energy. That is one reason polyethylene was so important

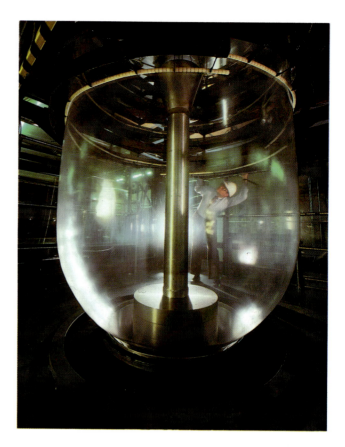

Polyethylene film, which is widely used for packaging, is formed by extruding the molten plastic through a ringlike gap and inflating it like a balloon.

to the development of radar, for an insulator was needed for cables carrying high-frequency alternating current; it is also why polyethylene is still widely used as an insulator.

The hydrocarbon character of the interior of a lump of polyethylene makes it a congenial home for other hydrocarbonlike molecules. Hence, polyethylene is a solvent for fats, oils, and grease; but since the polymer molecules are not very mobile, dissolving occurs very slowly, especially in the high-density polymers. Nevertheless, polyethylene articles do slowly absorb grease, become stained by it, and lose some of their electrical insulating qualities.

POLYPROPYLENE　(49)　$[CH(CH_3)CH_2]_n$

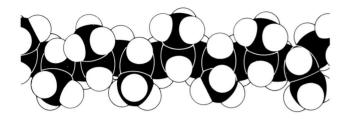

Many polymers are variations on the theme of polyethylene, differing only in having one or more of the hydrogen atoms in the monomer molecule replaced by other atoms or groups. The propylene molecule is an ethylene molecule in which a —CH_3 group has replaced one hydrogen atom. It is obtained along with ethylene when hydrocarbons are cracked, and it can be polymerized to form long molecules with a polyethylenelike backbone and —CH_3 groups on alternate carbon atoms. Special *catalysts are used to ensure that there is little chain branching and that the —CH_3 groups all point in the

same direction. Such an orderly polymer is said to be *isotactic,* and most of the polypropylene available commercially is of this kind.

The molecules in an isotactic polymer can lie close together, giving an extremely orderly solid. Because of its orderliness (its *crystallinity*), polypropylene is stiff, hard, and resistant to abrasion and has a high enough melting point for objects made from it to be sterilized. However, because the CH_3 groups are liable to *oxidation, polypropylene articles usually have *antioxidants (42) incorporated into them to divert attack by oxygen.

VINYL CHLORIDE　(50)　C_2H_3Cl
POLY(VINYL CHLORIDE)　(51)　$(CHClCH_2)_n$

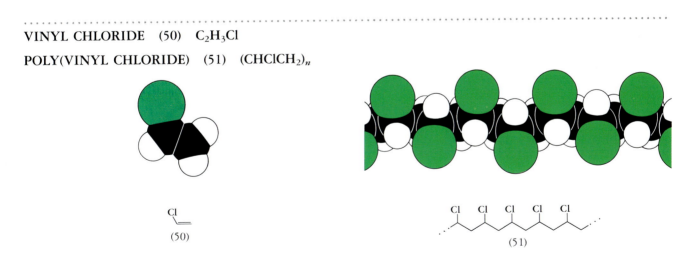

(50)

(51)

The vinyl chloride molecule can be considered as derived from an ethylene molecule by replacing a hydrogen atom with a chlorine atom (the green sphere in the illustration).

Vinyl chloride is a carcinogenic gas. Nevertheless, it is manufactured (from ethylene) in huge amounts each year, since it can be polymerized to form poly(vinyl chloride), or PVC, one of the most useful and adaptable of all plastics. Because of the big chlorine atoms present on the chains, PVC molecules do not pack together so well that they form a rigid solid. The usual flexible PVC contains large organic molecules (including *esters of alcohols containing about ten carbon atoms) that act as *plasticizers,* lubricating the PVC molecules so they will move easily past each other when the solid is bent. These lubricants are not bound chemically to the polymer chains, and they slowly migrate to the surface and are lost; this aging leaves the plastic brittle and stiff. Since some bacteria enjoy a good meal of hydrocarbon chains and will eat parts of the plasticizer molecules, biocides are sometimes incorporated into PVC.

PVC is produced in such huge amounts because it is so versatile. It can be mixed with a very wide range of additives chosen to tailor its properties to many different applications. When properly protected by its additives, it is also chemically resistant to attack and degradation.

Poly(vinyl chloride), the material from which these pipes are made, is one of the most widely used plastics. About 6 billion pounds are used annually in the United States, principally in the construction industry. About 25 million pounds are used for making credit cards.

VINYLIDENE CHLORIDE (52) $C_2H_2Cl_2$

POLY(VINYLIDENE CHLORIDE) (53) $(CCl_2CH_2)_n$

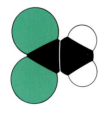

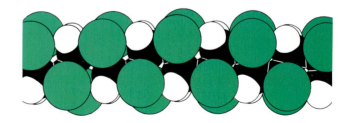

(52)

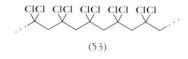

(53)

Vinylidene chloride may be polymerized on its own or copolymerized with vinyl chloride to give the polymers known collectively as *saran*. Saran A is poly(vinylidene chloride), the polymer obtained from vinylidene chloride alone; it consists of long chains in which —CCl_2— groups alternate with —CH_2— groups, as shown here. There are few —CCl_2—CCl_2— and —CH_2—CH_2— groups along the chain. Saran B, the more common variety, is the copolymer with vinyl chloride.

Poly(vinylidene chloride) molecules have a very regular structure, which is preserved to some extent in saran B. As a result, the molecules can pack together closely, giving a dense, high-melting-point substance. One consequence of this dense packing is the high impermeability of saran films to gases, which results in their use for cling wrapping. A second consequence is their low solubility in and impermeability to organic liquids—one reason why saran is so widely used for automobile seat covers.

TETRAFLUOROETHYLENE (54) C_2F_4

POLYTETRAFLUOROETHYLENE (55) $(CF_2CF_2)_n$

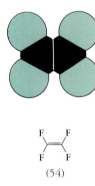

(54)

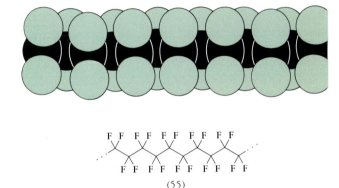

(55)

Tetrafluoroethylene is an example of a *fluorocarbon,* a compound of fluorine and carbon. In the molecule, all four hydrogen atoms of ethylene (47) have been replaced with fluorine atoms.

Fluorocarbon compounds constitute almost another world from organic and inorganic chemistry: the C—F bond is so strong that it almost totally ignores attack by other molecules. In the small fluorine atoms, the highly charged fluorine nucleus exerts such tight control over the electrons in its vicinity that they take much less part in weak intermolecular bonding than the electrons in hydrocarbons. As a result, fluorocarbons are generally more volatile than the corresponding hydrocarbons.

Fluorocarbons came into prominence after World War II, when the nuclear industries grew and supplies of fluorine became available [fluorine is used in the manufacture of uranium hexafluoride (UF_6), a volatile solid used in the separation of uranium isotopes]. Tetrafluoroethylene is a colorless, odorless, tasteless gas; its principal destination is polymerization to give the fluorocarbon analogue of polyethylene—polytetrafluoroethylene, or PTFE.

PTFE consists of very long chains, composed of about 50,000 —CF_2— groups each, with very little cross-linking between them. As a result, the molecules pack together to give a dense, compact solid with a high melting point. Even when the material is molten, the chains are so closely packed that they flow past each other only very slowly. Molten PTFE is so viscous that most PTFE articles are made by heating and compressing the powder to obtain a dense, strong, homogeneous lump.

The chemical and thermal stability of PTFE can be traced to two features. One is the considerable strength of the C—C and C—F bonds, which keeps the molecules from decomposing even when moderately heated. The second feature is the match between the sizes of the fluorine and carbon atoms, which results in the fluorine atoms forming an almost continuous sheath around the carbon atom chain, protecting it from chemical attack. In effect, the fluorine atoms act as chemical insulation around the carbon-atom "wire."

Grease and oil do not form bonds with PTFE, so surfaces coated in it are "nonstick" (that is, PTFE is an *abherent* coating, the opposite of an adherent substance like glue). Because fats and oils do not form bonds with the alien PTFE molecules, PTFE feels slippery to the touch. Its molecules pack together so densely that the solid does not absorb water, and hence it is an excellent electrical insulator.

Tetrafluorethylene is often copolymerized with other fluorocarbons to produce the range of plastics widely known as *Teflons* (a Du Pont trade name). One Teflon is PTFE itself. Another, Teflon FEP (the initials stand for fluorinated ethylene-propylene), is a copolymer of tetrafluoroethylene and the fully fluorinated version of propylene (CF_3—CF=CF_2). The —CF_3 groups of the fluoropropylene molecule are bumps on the —CF_2— backbone of the copolymer and result in less close packing. As may be suspected, the solid melts at a lower temperature than PTFE and forms a liquid with a lower viscosity. It can therefore be molded by conventional (injection) techniques but retains the desirable thermal and chemical stability of PTFE.

STYRENE (56) C_8H_8

POLYSTYRENE (57) $[CH(C_6H_5)CH_2]_n$

(56)

(57)

We now return to less extreme, still organic, modifications of ethylene, for many changes can be made by attaching different groups to the $CH_2{=}CH_2$ unit. A styrene molecule, for example, is obtained from an ethylene molecule by replacing one of the latter's hydrogen atoms with a benzene ring (more precisely, with the *phenyl* group ($-C_6H_5$), a benzene ring with one hydrogen atom removed). The benzene ring brings enough electrons to the molecule—and with them stronger intermolecular interactions—to convert ethylene, a gas, to styrene, a colorless liquid. Its principal destination is polymerization to polystyrene.

Styrene polymerizes to give long chains of $-CH(C_6H_5)-CH_2-$ units with very little cross-linking between them. Although *isotactic chains (in which the benzene rings are all on one side) can be prepared, the material that results is too brittle for general use. Most polystyrene consists of *atactic* molecules, in which the benzene rings point in random directions.

Because of the strength of the interactions between benzene groups on the chains, and because these groups obstruct the movement of chains past each other, polystyrene is less flexible than polyethylene. Flexible polystyrene can, however, be obtained by incorporating molecules that act as lubricants, as is done for PVC (51). The highly knobby character of its polymer chains and the random, disorderly way in which they pack together account for the great transparency of pure polystyrene. Lucite (59) has a similar transparency for a similar reason.

The benzene rings make polystyrene susceptible to damage by ultraviolet light and other high-energy forms of radiation, so *antioxidants are generally mixed in with polystyrene; they are especially needed in polystyrene fluorescent-light fixtures, for fluorescent lamps generate some ultraviolet light (page 16). There is enough energy in sunlight to induce radiation damage, and unless polystyrene is protected by antioxidants, it rapidly yellows and degrades. The color may be due to light absorption by an oxygen molecule lying close to and loosely bonded to one of the benzene rings.

AZODICARBONAMIDE (58) $C_2H_4O_2N_4$

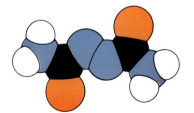

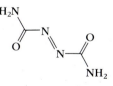

This small but action-packed molecule can be thought of as being derived from a nitrogen (N_2) molecule by opening one of its three bonds, attaching a carbon monoxide (CO) molecule to each fragment with a C—N bond, and completing each carbon's bonding tendency with an $-NH_2$ group, which is a fragment of ammonia (7).

Foamed polystyrene is prepared by adding a foaming agent, of which azodicarbonamide is an example, to the molten plastic. When it is heated, the azodicarbonamide molecule falls apart into the gases carbon monoxide, nitrogen, and ammonia and is captured as bubbles in the molten polymer, like a frozen head of beer.

Styrene is foamed using either a substance that decomposes to give rise to a gas or by incorporating a volatile liquid (such as hexane) in the polymerization mixture and subsequently heating it. The vaporized liquid forms bubbles in the softened polymer.

METHYL METHACRYLATE   (59) $C_5H_8O_2$

Methyl methacrylate is the monomer from which poly-(methyl methacrylate) is made, and poly(methyl methacrylate) is better known as *Lucite, Plexiglass,* or *Perspex.* Bulky, irregular side groups attached to the basic ethylene fragment cause the polymer chains to lie together in a very irregular way, so that the solid is internally very chaotic. This is just like the lack of arrangement of water molecules in a glass of water. Since the solid is amorphous on a molecular scale, it does not scatter light that passes through. Consequently, blocks of the polymer are brilliantly transparent, like clean water.

One of the side groups attached to the ethylene fragment is —CH$_3$. Its importance in contributing to the knobbiness of the polymer, and hence to its rigidity, can

be seen by removing it from the monomer, which gives methyl acrylate. This latter ester can be polymerized in the presence of a *surfactant; the resulting *acrylic ester polymer* is in the form of minute droplets that remain suspended as a milky *emulsion. This substance is the base for acrylic paints, which are formed simply by adding pigment. When the water evaporates from the emulsion after it has been applied to a surface, the acrylic ester polymer remains as a rubbery, flexible film that retains the pigment. The polymer molecule has the additional advantage of not absorbing the ultraviolet light present in the sunlight that reaches the surface of the earth, so the paint does not degrade, break up, or, so long as the pigment survives, lose its color.

Self-Portrait, 1966, by Andy Warhol. Synthetic polymer paint and enamel silk-screened on six canvases.

LAURYL METHACRYLATE (60) $C_{16}H_{30}O_2$

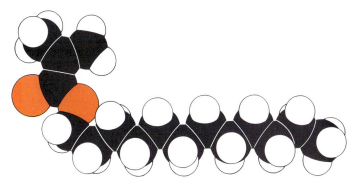

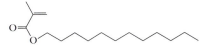

The lauryl methacrylate molecule is a modification of the methyl methacrylate molecule (59) in which the hydrocarbon tail has become a chain of 12 carbon atoms. Poly(lauryl methacrylate), the polymer formed from this monomer, and its close relatives are used as additives in viscostatic engine oils. The technological problem to overcome is the loss of viscosity (and the attendant draining from surfaces) that occurs when an oil is heated. An oil that is viscous when the engine is cold is not the answer, because such an oil can not be pumped over engine surfaces adequately when the engine is first started.

The long hydrocarbon tail of poly(lauryl methacrylate) molecules makes them soluble in the similar environment that exists inside lubricating oils, which consist of hydrocarbons containing more than about 12 carbon atoms. (The thicker the oil, the longer the hydrocarbon chains.) At low temperatures, the polymer molecules are coiled into balls, so they do not hinder the flow of the hydrocarbon oil molecules very much. As the temperature is raised, the coils unwind because the atoms in their hydrocarbon chains move more vigorously and flay through their surroundings. The unwound polymer chains stretch through a greater region of the oil and hinder the hydrocarbon oil molecules much more; thus the oil flows more slowly (is more viscous) than it would if the additive were absent.

....

METHYL CYANOACRYLATE (61) $C_5H_5O_2N$

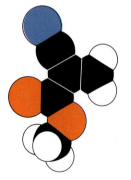

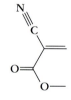

Methyl cyanoacrylate is the substance contained in tubes of *Super Glue*. When it is spread on surfaces that are to be joined, it begins to polymerize because the surfaces contain traces of water and alcohols. This readiness to polymerize is due to the presence of two very strongly electron-attracting groups that are attached to the same carbon atom: The *cyanide group* —C≡N withdraws electrons from the carbon it is attached to, as does the oxygen-rich carboxyl group on the same atom. This greatly distorts the electron distribution in the carbon-carbon double bond and makes it very susceptible to attack.

The formation of poly(methyl cyanoacrylate) proceeds quickly once it is initiated; the surfaces are joined partly because cavities and tiny crevices in them are filled, which locks them together, and partly because the polymer forms chemical bonds with them. The glue sticks hard to skin. This may be a disadvantage for individuals in normal health, but it is turned to advantage as illness advances: Surgeons can use Super Glue in place of sutures, and morticians use it to seal, once and for all, the eyes and lips of their clients.

RUBBER

Rubber, so called (by Joseph Priestley) because it can be used to rub out pencil marks, is an example of an *elastomer,* an elastic polymer. Natural rubber is obtained commercially from the coagulated latex of *Hevea brasiliensis,* but it is also present in the hollow stem of the dandelion (*Taraxacum officinale*). It is hard and brittle when cold, and sticky when warm. Its elastic properties are much improved if it is first shaped into the desired form and then heated with sulfur. This is the process of *vulcanization,* invented by Charles Goodyear in 1839.

The commercial source of natural rubber is the rubber tree, *Hevea brasiliensis.* The white edge on the leaf is exuding latex.

ISOPRENE   (62)   C$_5$H$_8$

POLYISOPRENE   (63)   [CH$_2$C(CH$_3$)=CHCH$_2$]$_n$

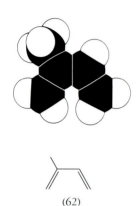

(62)

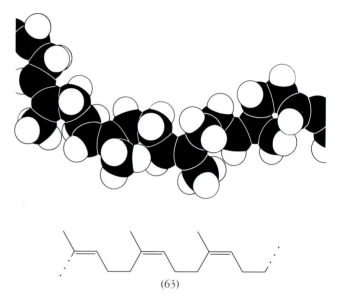

(63)

Isoprene, a volatile, fragrant liquid, is the *monomer that *polymerizes to form natural rubber. Rubber molecules consist of chains of several thousand —CH$_2$—C(CH$_3$)=CHCH$_2$— units. The presence of a double bond in each unit is an important feature, as you will see shortly.

Hot rubber smells of isoprene, for when it is heated some isoprene molecules break loose. [People also smell slightly of isoprene, for a different reason (page 140).] This was the first clue that isoprene was a building block of rubber, yet early attempts to synthesize rubber from isoprene failed. This failure was eventually traced to the fact that when isoprene molecules link together, they can do so in either of two different ways. In natural rubber, all the linking between the units is done by enzymes working under the dictates of genetic control, and *H. brasiliensis* produces a polymer with all links in the cis arrangement. Early attempts to synthesize the polymer had produced chains with a random mixture of cis and trans units, and the material was sticky and useless. This problem has since been overcome by using a suitable *catalyst (in this case, a compound that includes aluminum and titanium atoms), and nearly pure *cis*-polyisoprene, with properties very similar to natural rubber, can be produced.

Gutta-percha, the material used to cover golf balls, is the version of polyisoprene in which all the units are trans. It is obtained by boiling the sap of species of trees of the order Sapotaceae, found in South East Asia. It is a much harder material than rubber and more resistant to water.

In unstretched rubber, the long molecules lie in tangled coils. Stretching the rubber stretches out the coils. When the stretched rubber is released, it will spring back to its original shape if the polymer molecules have not slipped past each other. In natural rubber, the stretched coils do slide past each other under stress, so natural rubber does not resume its original shape exactly.

The problem of molecule slippage is overcome by vulcanization. When natural rubber is heated with sulfur, the sulfur attacks the double bonds that remain in the chains and forms —S—S— bridges between neighboring molecules. This improves the elastic properties of the solid, because now the molecules are linked into a huge three-dimensional network. Neighboring molecules can no longer be moved apart, but they are still moderately free to unwind under stress as long as neighbors cooperate. Hence the solid is more resilient, and it recovers its shape when the stress is removed. Human hair (77) is naturally slightly vulcanized and so is springy for similar reasons.

Heavily vulcanized rubber is the hard, tough, and chemically resistant material called *ebonite* or *vulcanite*.

....

ISOBUTYLENE (64) C_4H_8

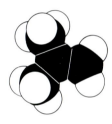

Chemists have tinkered with the composition of rubber and have produced a variety of elastomers. *Butyl rubber* is obtained through the copolymerization of isobutylene (which is more formally called *2-methylpropene*) with a little isoprene. Since isobutylene has only one double bond, which is used in the polymerization process, butyl rubber has fewer double bonds than natural rubber. It still has enough so that it can be vulcanized, but it withstands weathering better than other elastomers because its molecules have fewer places where atmospheric oxygen can attack.

The molecules of butyl rubber pack more closely than those in natural rubber, so that it is much less permeable to gases. It is used for inner tubes and for the interiors of tubeless tires.

Another starting point for new types of rubber is an isoprene molecule in which a —CH$_3$ group has been replaced by a chlorine atom. When this *chlorobutadiene* is polymerized, it gives a structure like that of natural rubber, but with each —CH$_3$ group replaced by a chlorine atom. This "chloroprene" elastomer is called *neoprene*. It is unusual in that it can be vulcanized by heat alone, with the molecules forming carbon-carbon links directly to each other without the need for bridges built from sulfur. It is resistant to oxidation, oil, and heat and is widely used in automobile parts.

Numerous other elastomers can be tailored to suit a particular task through judicious choices of monomers and of the type of copolymerization. One very widely used composition is *styrene-butadiene rubber,* or SBR, the copolymer obtained from a mixture of styrene (56) and butadiene. Most of this product is vulcanized and used for automobile tires. A minor use for the raw (unvulcanized) material is as chewing gum (117).

POLYESTERS AND ACRYLICS

Artificial (as well as natural) fibers should consist of long molecules that can be made to lie parallel to each other as they are drawn out into a thread. One of the ways of achieving this is to link together an acid and an alcohol molecule as in the process of *ester formation, but in such a way that the ester molecule can go on growing at each end. This results in indefinitely long molecules of repeating units called *polyesters*.

ETHYLENE GLYCOL   (65)   C₂H₆O₂

The presence of —OH groups at both ends of the ethylene glycol molecule has important physical and chemical consequences. Since ethylene glycol can form *hydrogen bonds at each end, whereas ethanol (27) can do so only at one end, we can expect it to be a more viscous and less volatile liquid. This is the case. It is also fully miscible with water, with which it can form hydrogen bonds. Less easy to predict is its toxicity, for a dose of little more than 50 milliliters can kill.

Ethylene glycol is the alcohol used as *antifreeze* for automobile engines. When it mixes with water, it attracts water molecules strongly and interferes with the interactions between them. As a result, they do not crystallize into a solid until the temperature has been lowered to well below the normal freezing point. In a sense, the ethylene glycol molecule acts as a hydrocarbon lubricant for the water molecules, which keeps the fluid supple. The glycol's low volatility is also an advantage, for it does not evaporate from the hot coolant.

The principal chemical importance of the ethylene glycol molecule lies in its Janus-like double alcohol character. Since it has an —OH group at each end, it can link with two carboxylic acids. That is, it can be converted to an ester at each end. Why this is important you will see shortly.

TEREPHTHALIC ACID   (66)   C₈H₆O₄

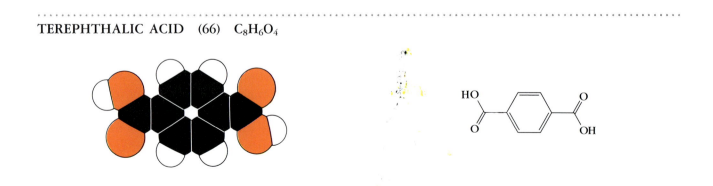

The terephthalic acid molecule is derived, both structurally and in practice, from the para-xylene molecule (25) by *oxidation of the two diametrically opposite —CH$_3$ groups. Terephthalic acid is a white, crystalline solid. Benzene (23) alone is a liquid, but the carboxyl groups in terephthalic acid, with their ability to form hydrogen bonds, can bind neighboring molecules together into a solid. Its chemically important property is its two-faced acid character, for, echoing ethylene glycol, it can form an ester at each end.

What, then, will result when this Janus-like acid is esterified with that Janus-like alcohol?

POLY(ETHYLENE TEREPHTHALATE) (67) $(O_2CC_6H_4CO_2C_2H_4)_n$

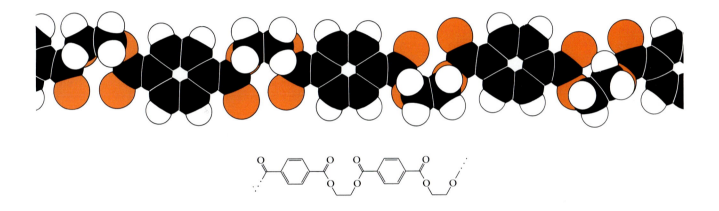

When terephthalic acid is allowed to react with ethylene glycol, long chains of polyester grow. First one glycol —OH group combines with one of the acid carboxyl groups, linking the two molecules together. But this still leaves the glycol with one free —OH group that can react with another carboxyl group and the acid with a free carboxyl group that can react with another glycol. Even after the next round of reaction, there are still an —OH and a carboxyl group at opposite ends of the molecule, and it remains ripe for further reaction. This process stops only when all the reagents have been consumed, or a chain combines with its other end, or an impurity with only one —OH or one carboxyl group ends the chain. The product, poly(ethylene terephthalate) is known as *Dacron* in the United States, *Terylene* and *Crimplene* in the United Kingdom, and *Trevira* in Germany.

The molten polymer is extruded through a spinneret, and the resulting fiber is then stretched to several times its original length. This uncoils the PET molecules, and they lie together in an orderly arrangement. The alternating benzene rings stiffen the chains and raise the melting point above that of polymers made up of simple flexible hydrocarbon chains. The stiffness is responsible for the crease resistance of no-iron fabrics made from the fibers. Thin film made from the polymer is sold under the trade name *Mylar* and used for cassette tape; the orderly orientation of the molecules in the film—which

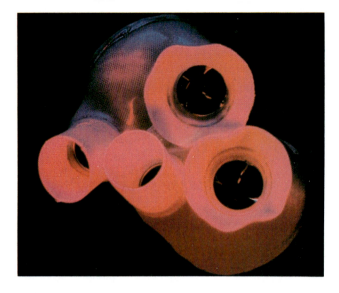

The Jarvik heart, the first synthetic heart, is made of Dacron. The blood enters by the upper pair of tubes and leaves by the lower. Because the polymer has a different texture from natural muscle it can affect the flow of blood, which can lead to an increased risk of thrombosis.

leads to a strong, almost unstretchable product—is achieved by stretching it in one direction.

Other varieties of polyesters are obtained by polymerizing other carboxylic acids with other alcohols. The straightness of the polyester molecules, and hence their fiber-forming character, can be eliminated by choosing an acid that introduces bends. One such is *phthalic acid,* which is like terephthalic acid but has the carboxyl groups as neighbors on the benzene ring. Cross-linking among the chains is achieved if, instead of ethylene glycol, glycerol (34) is used (or included), for its third —OH group can begin a branch line. The polymers that result from such monomers are called *alkyd resins,* and they form flexible sheets rather than fibers. They are available as *emulsions in water and are widely used as surface coatings, including paints.

....

ACRYLONITRILE (68) C_3H_3N

POLYACRYLONITRILE (69) $[CH(CN)CH_2]_n$

(68)

(69)

Many people unknowingly spread cyanide over their bodies, for they dress themselves in *acrylics,* polymers obtained by using acrylonitrile as a monomer. The polymerization process is like that for polyethylene (48), and the molecules link together to produce a polyethylenelike chain called polyacrylonitrile, in which every other carbon atom carries a cyanide group. The polymer, which is sold as *Orlon,* is dissolved in a solvent, and then fibers are spun by squirting the solution through spinnerets into air so that the solvent evaporates. The fibers are then stretched to several times their original length to orient the molecules so that they lie parallel to each other.

Pure polyacrylonitrile resists dyeing, so it is common to introduce small amounts of other monomers, including styrene (56), vinyl chloride (50), and vinylidene chloride (52), to give a copolymer with points of attachment for dyes. These copolymers are called *modacrylics.* Copolymerization with vinyl acetate, a molecule in which a

group takes the place of the cyanide group, gives *Acrilan.*

The presence of the chlorine atoms in the vinyl chloride and vinylidine modacrylics results in improved flame resistance; most acrylics used for carpets (and for the personal analogue of carpets—wigs) are actually modacrylics with a substantial chlorine (or bromine) content. Flame resistance is a very complex property to which many features contribute. These include achieving a higher ignition temperature of parts of the polymer chain and introducing weakness into the polymer, so that the fabric drips away before a flame can spread. The combustion products may also smother the fabric and thus help to exclude air. They may also inhibit the spread of a flame by introducing radicals which interfere with the *chain reaction that is fire.

Acrylics are resistant to digestion by microorganisms in the soil and to degradation by sunlight. Hence they are suitable for use out of doors, and artificial grass is sometimes an acrylic.

NYLON

Nylon, the first completely synthetic fiber, was developed in the mid-1930s at the laboratories of Du Pont. It still retains a very large share of its market.

Nylon is an example of a *polyamide,* a polymer that resembles a polyester (67) but results from the reaction between —NH$_2$ (rather than —OH) and a carboxyl group. This difference has important consequences. For example, the extra hydrogen atom in —NH$_2$, which survives after the *amide link is formed (in contrast to the single hydrogen atom of —OH, which is removed), provides the opportunity for *hydrogen bonding. And with hydrogen bonding, as you have seen, comes strength.

There are numerous polyamides, and they are named according to the number of carbon atoms in the monomers. The two most important, in the sense of taking the lion's share of the market, are nylon-6,6 and nylon-6. For historical reasons that partly reflect the availability of raw materials, nylon-6,6 originally predominated in the United States, and nylon-6 in Europe. Polyamides in which the monomers are derived from benzene (23), so that they have chains of hexagonal benzene rings joined by amide linkages, are generically known as *aramids* (a word obtained from the fusion of "aromatic," a general term for derivatives of benzene, and "amide").

....

ADIPIC ACID   (70)   C$_6$H$_{10}$O$_4$

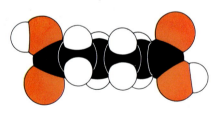

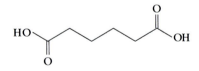

Adipic acid is a white crystalline solid. It resembles terephthalic acid (66) in that it has an acid group at each end; like terephthalic acid, it can also grow at each end. If it is esterified with a double alcohol such as ethylene glycol (65), it will grow into a polyester. You will soon see how it can grow into a polyamide, and how its six carbon atoms contribute one of the sixes to nylon-6,6.

The two carboxyl groups of adipic acid have just enough *hydrogen bonding capacity to overcome the hydrocarbon character of the —CH$_2$— chain and to drag it into solution in water, in which it is slightly soluble. It is an approved food additive, and it has been used to acidify soft drinks. Adipic acid also contributes to the sharp taste of beets (*Beta vulgaris*).

HEXAMETHYLENEDIAMINE (71) $C_6H_{16}N_2$

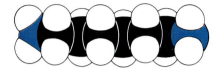

Each —NH_2 group in this molecule is called an *amino group*. The amino group is the nitrogen analogue of the —OH group, and amines are to some extent the nitrogen analogues of alcohols. However, as always in chemistry, the replacement of one atom with another often brings in its trail profound consequences, even though the two atoms may be quite similar. The step from the world of oxygen chemistry to that of nitrogen is accompanied by two important changes. One is the presence of an additional hydrogen atom in —NH_2, as compared

with —OH. The second is a more subtle change: Nitrogen has a weaker nuclear charge than oxygen and hence has slightly less control over its electrons. This apparently innocuous difference effloresces into the world with great consequences for nylon and for life, as you will see next.

Hexamethylene diamine is closely related to some evil-smelling amines. To discover how narrow the line is between your clothing and your putrefaction, see *putrescine* (131).

POLY(HEXAMETHYLENE ADIPAMIDE) (72) $[CO(CH_2)_4CONH(CH_2)_6NH]_n$

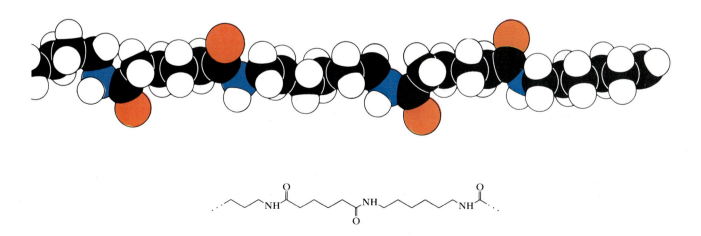

In the industrial version of a spider's spinneret (see pages 93–94), nylon is being extruded as thread.

A polymer molecule is formed when a diamine such as hexamethylenediamine is allowed to react with a dicarboxylic acid like adipic acid. The two react to produce an *amide,* a compound characterized by the group of atoms

$$\begin{array}{cc} O & H \\ \parallel & \mid \\ -C & -N- \end{array}$$

This group is often written —CO—NH—. As in polyester formation (67) their ability to react is not quenched, because an amino group and a carboxyl group remain at opposite ends of the new molecule; growth can continue there by reaction with other dicarboxylic acid and diamine molecules. Hence, the chain can grow indefinitely, to produce a very long molecule (only a fragment of which is shown).

One manufacturing problem, that of mixing the diamine and the acid in equal proportions, can be solved elegantly by making use of nitrogen's lesser control over its electrons: Its *lone pair of electrons can accept a hydrogen nucleus from the acidic carboxyl group. Thus, when adipic acid and hexamethylenediamine are mixed in nearly equal proportions, the double acid and the double amine form a double salt, commonly known as *nylon salt,* with —CO_2^- groups as anions and —NH_3^+ groups as cations. Any excess acid or diamine is then removed, which ensures that there is one diamine molecule for every acid molecule. When the salt is heated, the —CO_2^- groups of the acid react with the —NH_3^+ groups of a neighboring diamine, and long chains of polyamide are formed. These chains can be thought of as polyethylene (48) molecules that are interrupted after every few —CH_2— groups by an amide group. As you will see, the differences between polyethylene and polyamides stem from the presence of the electron-hungry oxygen and (to a lesser extent) nitrogen atoms that are strung along the chain.

The result of polymerizing the six-carbon-atom hexamethylenediamine with the six-carbon-atom adipic acid is *nylon-6,6,* one of the most important of the polyamides. It is extruded from the reaction chamber and cut into cubes for molding or else spun from the melt for fiber. The spinning orients the polyamide molecules parallel to each other. In that position the chains can stick together by forming N—H···O hydrogen bonds between the NH groups of one chain and the CO groups of another. This *hydrogen bonding gives the fibers their great strength. It also accounts for the good elastic recovery of nylon, for the hydrogen bonds act like the sulfur bridges in vulcanized rubber (page 79): They pull the fibers back to their original arrangement when stress is removed. Nylon stockings hug moving legs by virtue of the hydrogen bonds between their molecules; polyethylene stockings would just sag.

Nylon-6,6 is a strong, tough, abrasion-resistant material with moderate water resistance. It is less water-resistant than a pure hydrocarbon polymer such as polyethylene (48), because water molecules can worm their way in by latching onto the amide groups by means of hydrogen bonding. Although nylon-6,6 is satisfactory as an insulation for wires carrying low-frequency electric current, its electrical properties are not as good as polyethylene's at and above radio frequencies. High-frequency electric fields make the >C=O and >N—H groups (which are

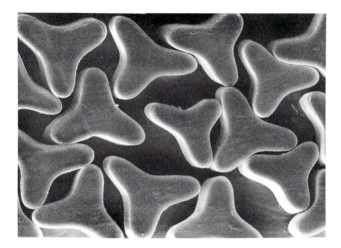

A cross section of nylon-6,6 staple (magnified 240 times) used for carpet weave. Walking on a carpet is similar to walking on a synthetic protein.

absent from polyethylene) waggle, hence the polymer molecules start to vibrate. This absorbs energy, and the electric signal is attenuated more rapidly than if polyethylene were used. The presence of the electron-attracting oxygen and nitrogen atoms in the polymer molecules enables a fabric made of nylon to pick up an electric charge when surfaces rub together. This is the origin of the mild electric shock that we sometimes experience after walking across a nylon carpet in a dry atmosphere.

Other polyamides are made with monomers containing different numbers of carbon atoms. As a rule, the longer the carbon chains, the more water-resistant the polymer. This is because a longer chain makes the polyamide more closely resemble a pool of oil, and water is more strongly repelled.

As noted in the introduction to this section, aramids are polyamides obtained by use of monomers derived from benzene [such as terephthalic acid (66) in place of adipic acid]. An example is the strong, rigid, and less water-absorbing substance marketed as *Kevlar*. The aramid chains can stack together closely as a result of both the hydrogen bonds between their amide groups and interactions between benzene rings, which can lie face to face. This gives a highly crystalline and very strong material that is used for tire cords and bulletproof vests.

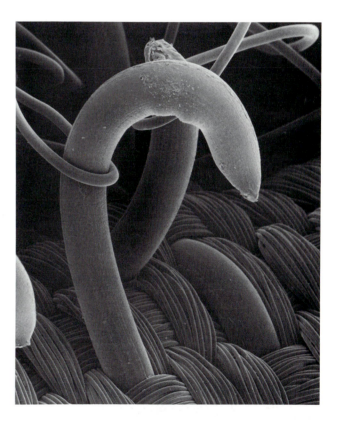

A scanning electron micrograph of Velcro being pulled apart. One half of the fastener is a surface with nylon loops; the other is a surface with hooks.

HAIR, WOOL, AND SILK

Hair, wool, and silk are *polypeptides*, compounds in which the characteristic feature is the repeating unit —CO—NH—C—, with various groups dangling from the second carbon atom. They are nature's version of nylon (72), the principal difference being that there is only one carbon atom between each pair of —CO—NH— units, rather than the half dozen or so typical of polyamides. Nevertheless, a wide variety of groups can be attached to that carbon, and polypeptides are much more varied than the nylons that have been synthesized so far. Indeed, wool and silk are examples of *proteins*. Proteins (from the Greek word for "primary") are the building blocks of living things, and they include the *enzymes, the worker-molecules of the hive of cells in our bodies. The building blocks of the proteins, in turn are the twenty *amino acids* that occur naturally. We examine one or two of them in this section.

Humans can synthesize about a dozen of the twenty amino acids, but about eight must be ingested in the diet: These are the *essential amino acids*. Because we are so similar to animals, eating their flesh gives a ready supply of all the amino acids we need. Plants have trodden a different evolutionary trail from our common ancestor, and we cannot be confident that all the amino acids we need will be present in a single vegetable source. Cereals, for instance, are generally deficient in one amino acid, lysine. Cultures in which meat is either too great a luxury or a moral abhorrence can circumvent such deficiencies by balancing a vegetable thin in a particular amino acid with one rich in it; in some cases there is a further complementarity, for the first vegetable may be rich in an amino acid that is at least partially absent in the second. This is possibly the origin of characteristic ethnic culinary preparations, such as the soybean and rice combination typical of the orient, the bean and corn combinations of Central America, and the macaroni and cheese combination of Italy.

Some amino acids contain sulfur. Indeed, almost all the 150 grams of that element present in a human body is in the amino acids that are found predominantly in the proteins of the hair, skin, and nails. Brazil nuts (the seeds of the Amazon tree *Bertholletia excelsa*) are a rich source of these proteins, but I am not aware that they have been tested as a cure for baldness or that their oil is used (as a gimmick) in shampoos.

....

GLYCINE (73) $C_2H_5O_2N$

The glycine molecule can be considered as a molecule of acetic acid (32) in which one of the hydrogen atoms of the —CH$_3$ group has been replaced by —NH$_2$, the *amino group*. All the amino acids in proteins are built on glycine's pattern, with the amino group attached to the carbon atom immediately adjacent to the carboxyl group. That is, they are all *α-amino acids* (as distinct from β-, γ-, and so on, in which the amino group is progressively further from the carboxyl group).

A glycine molecule thus possesses a carboxyl group (which makes it an acid) next to an amino group. It is this juxtaposition of these two reactive groups that makes α-amino acids, of which glycine is the simplest, so adaptable and important. Glycine units can be strung together in a polyamide chain; the result is *polyglycine*, (—NH—CH$_2$—CO—)$_n$, the world's most boring protein. But although it is itself not particularly interesting, polyglycine is the backbone that supports life, as you will now see.

··

ALANINE (74) C$_3$H$_7$O$_2$N

VALINE (75) C$_5$H$_{11}$O$_2$N

(74L)   (74D)

(75)

The possibilities for variety in proteins get richer as soon as amino acids are obtained in which different groups are attached to the spare carbon atom. The simplest possibility is to introduce a —CH$_2$— group between a carbon atom and a hydrogen atom of glycine, obtaining alanine. Alanine is the nitrogen analogue of lactic acid (33), which has an —OH group in place of alanine's —NH$_2$.

If the process of inserting —CH$_2$— is allowed to continue, various other amino acid molecules can be created, especially if the hydrocarbon tail is allowed to branch. One of them is valine, which has a bulkier and oilier side chain than alanine (a point we shall return to). Only about twenty such substitutions are found in nature, and all the proteins in our bodies are built from permutations of them in a polypeptide string.

An interesting detail of nature is that although two versions of each amino acid (other than glycine) are possible, only one is found naturally. The two versions of alanine, L-alanine and D-alanine, are shown as (74L) and (74D), respectively. They differ like a left hand (L) and a right hand (D), in that one is the mirror image of the other. The same is true of lactic acid (33); however, unlike lactic acid, all naturally occurring amino acids on earth, and hence all proteins, are built from only *left-handed* amino acids. Thus life, in a sense, is fundamentally left-handed. The reason for this left-handedness is not known, but some people speculate that it may be connected with a similar asymmetry in the properties of elementary particles.

POLYPEPTIDES (76)

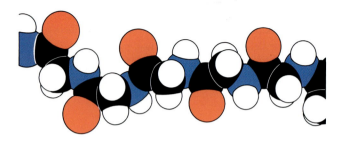

The linking of amino acids into a polypeptide chain occurs when an amino group of one molecule reacts with a
carboxyl group of a neighbor, exactly as in the production of a polyamide (72). However, with twenty different amino acids available, polypeptides can be formed
with almost infinite variety. The illustration on the right
shows one of the four chains that link together to give
the protein *hemoglobin*, the molecule responsible for
transporting oxygen in the blood. The oxygen molecule
attaches reversibly to the iron atom at the heart of the
molecule (provided its place has not already been occu-
pied by a carbon monoxide molecule as a result of smoking or a cyanide ion in certain other cases of poisoning).
The order in which amino acids are joined to give a peptide is determined ultimately by the genetic material
DNA in the nuclei of cells.

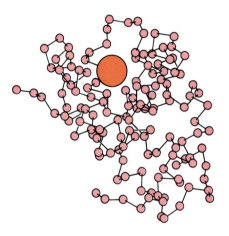

A model of one of the four polypeptide chains that make up a hemoglobin molecule. There is an iron atom at the center of each twisted
and folded chain (recall the pigs on page 18).

α-KERATIN (77)

Hair and *wool* consist of α-keratin. This protein is made
up of chains of amino acids in which glycine and leucine
figure prominantly, with about half a dozen others play-
ing an equally important role. Many of the acids have
bulky side groups, as does leucine, and some carry sulfur
atoms.

The polypeptide chain forms a right-handed helix,
called an *α-helix*, with a shape maintained by hydrogen
bonds between different amino acids. Three of these
right-handed helices wrap around each other in a left-
handed coil, where they are held together by more hydrogen bonds and some sulfur bridges. The sulfur
bridges reach between amino acids that contain sulfur
atoms and resemble the bridges in vulcanized rubber
(page 79). Nine of these coils cluster around two more,
giving a *microfibril* of eleven coils, each consisting of

three α-helices. Hundreds of these microfibrils are embedded in an amorphous protein matrix to give a *macro-fibril*, and these macrofibrils stack together to give a hair cell. A hair *fiber*, in turn, consists of a stack of these cells.

The extensibility of wool and hair is due to the ability of the highly wound structure to unwind, even as far down as to unwind the α-helices, when the hydrogen bonds that support it are broken. The shape is restored when the tension is released because the sulfur bridges survive the stretching (as they do in vulcanized rubber) and snap the polypeptide back into its helical arrangement. In the *permanent waving* of hair, the sulfur bridges are broken and the hair is stretched, after which the bridges reform in a more fashionable arrangement.

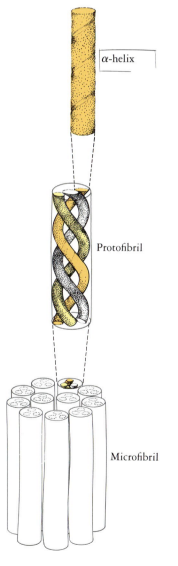

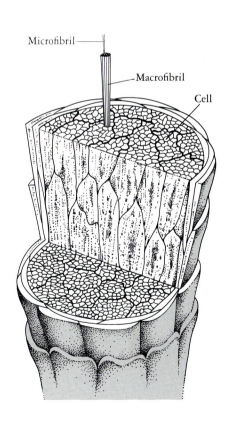

A strand of hair can be successively taken apart into smaller fibers, macrofibrils, microfibrils, and the keratin molecule itself.

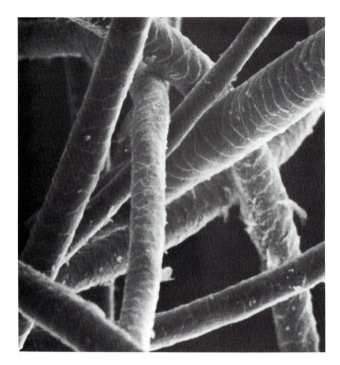

Wool fibers. The scales are responsible for the formation of felt, for they lock the fibers into a dense tangle when they are rubbed together.

Claws, nails, and hooves are also keratin, but they are more highly cross-linked with sulfur bridges (in a sense, more heavily vulcanized, as ebonite is) and are more rigid.

The color of black, brown, and fair hair is due to various concentrations of *melanin* (142). Red hair is colored by a pigment (trichosiderin) based on iron, like blood and rust. The bleaching of hair is usually an attack on the compounds responsible for its color; it is almost always accomplished with dilute solutions of hydrogen peroxide (14), which oxidizes the molecules. A side effect of hydrogen peroxide bleaching is that it leads to the formation of more sulfur bridges (by removing hydrogen atoms from some —SH groups, enabling the remaining —S atoms to form bonds). This increased "vulcanization" makes bleached hair more brittle. For those wishing to reclaim the appearance of youth, gray hair can be blackened with preparations that contain lead acetate.

Here, once again, sulfur is the pivot of a cosmetic affectation, for the lead ions combine with the sulfur of the amino acids, in effect forming black lead sulfide. The same blackening was less welcome when it affected white paint, which once contained lead, in industrial areas where sulfur compounds were present in the air (page 27).

The luster of hair is its ability to reflect light. Some alkaline hair preparations and shampoos remove hydrogen ions from the keratin molecules, thus altering their electric charge distribution. As a result, they and the microfibrils coil more tightly and become more reflective, enhancing the luster. *Hair conditioners* include ionic substances (organic derivatives of nitrogen) that attach to the fibers and modify their electric charge. This increases the electrical repulsion between hairs that happen to approach each other; since they cannot stick together, the hair is given a sense of having "body."

β-KERATIN (78)

Silk is the solidified fluid excreted by a number of insects and spiders, the most valuable being the exudent of the silkworm, the caterpillar of the silk moth (*Bombyx mori*). *Wild silk* is obtained from the night peacock moth (*Antheraea pernyi*) which does not feed on the mulberry and is not domesticated.

Silk, the common name for β-keratin, is a polypeptide made largely from glycine and alanine (and smaller amounts of other amino acids, principally serine and tyrosine). Most of the amino acids do not have the bulky side groups characteristic of those contained in the wool polypeptide. Partly as a result of their smaller side groups, β-keratin molecules do not form a helix; instead they lie on top of each other to give ridged sheets of linked amino acids, with the glycine appearing on only one side of the sheets. The sheets then stack one on top of the other.

This planar structure is felt when you touch the smooth surface of silk. Silk is less extensible than wool because its polypeptide chains are all nearly fully extended (as in pulled nylon fibers). However, it is flexible, because the sheets are only loosely bonded to each other (by *hydrogen bonds) and can slide over each other reasonably freely.

Silk is also the substance extruded by spiders: here the web is being spun from the silk extruded from the spider's spinnerets.

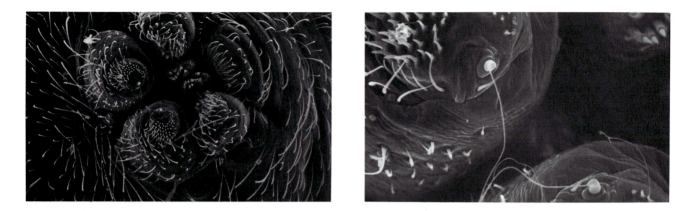

The spinnerets of a spider. The two on the left in the left-hand photo are magnified in the photo on the right, and the exuding silk can be seen.

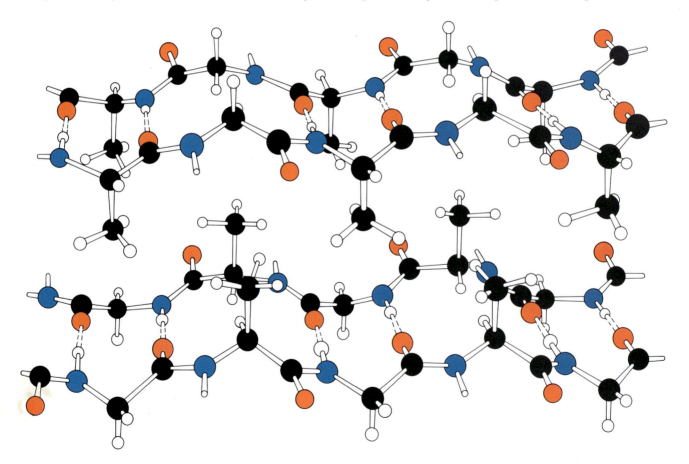

Silk consists of amino acid molecules linked into polypeptide chains. The chains themselves are linked by hydrogen bonds and lie in great puckered but almost flat sheets.

SUGAR, STARCH, AND CELLULOSE

Sugars, starch, and cellulose may seem strange bedfellows for nylon and wool. However, they continue the theme of this chapter, in which we have seen complexity being spun by the repetition of a single unit. Like proteins, starch and cellulose are natural polymers; the simple repeated unit is *glucose* or a similar molecule. Edible *starch* and inedible, structural *cellulose* (the most abundant organic chemical on earth) are examples of *carbohydrates*, substances with formulas that are often multiples of CH_2O, which is (falsely) suggestive of "hydrates of carbon." So surely and thoroughly does nature explore the opportunities open to its materials, that in *wood* it has already stumbled upon the cellulose analogue of foamed polystyrene.

GLUCOSE (79) $C_6H_{12}O_6$

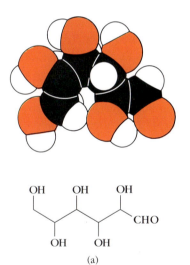

(a)

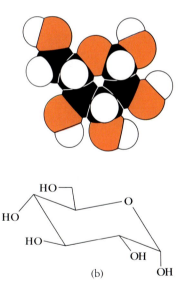

(b)

The glucose molecule exists in two main forms. In one (79a), it is a string of six carbon atoms; five of them carry —OH groups and the sixth has a doubly bonded oxygen atom, so that the line ends in a —CHO group. In the other (79b), a similar line of atoms has bent around and the —CHO group has reacted with an —OH group near the far end of the molecule to form a closed six-member ring, with one of the atoms in the ring being an oxygen atom. A solution of glucose should be pictured as consisting of a writhing, continuously interchanging collection of open molecules, six-member rings, and five-member rings like the one shown for fructose (80).

Some of the properties of glucose should be obvious from the structure of the molecule. In particular, the oxygen atoms bring it properties that are quite different from those of its parent hydrocarbon *hexane*, in which each of six carbon atoms carries only hydrogen atoms [as in octane (20)]. These numerous oxygen atoms result in glucose being highly soluble in water, since they can form strong *hydrogen bonds to water molecules. Note particularly that, in the six-member ring (79b), all the —OH groups are arranged along the perimeter like teeth on a gear wheel; this high degree of exposure allows water molecules to form numerous strong hydrogen bonds with them. Consequently, the molecule readily slips into solution.

Glucose resembles hexane in being a fuel, because its carbon atoms are ripe for conversion into carbon dioxide (4) by oxygen. But here we see a compromise: Hexane carries no oxygen atoms, and many new and strong carbon-oxygen bonds are formed when it burns. Glucose is already partly oxidized, and fewer new carbon-oxygen bonds are formed. However, hexane is not soluble in water, and so cannot be transported to cells by blood. Glucose on the other hand, can be regarded as a strip of water-soluble carbon atoms, less efficient perhaps than hexane but much more readily transportable through the body.

Glucose is also known as *dextrose* (because solutions of glucose rotate the plane of *polarized light to the right).

It occurs in ripe fruits, the nectar of flowers, leaves, saps, and blood and is variously called *starch sugar*, *blood sugar*, *grape sugar*, and *corn sugar*. It is the primary fuel for biological cells, and more complex sugars and starches are broken down into individual glucose units when they are digested. Once glucose molecules are in solution, their energy is immediately available for any demands that a metabolic process may make. The hydrocarbons stored in adipose tissue (in the obese) are available only much more slowly, since they must first be made soluble, then transported, and only then can be used.

Glucose is so soluble in water that it grips water molecules to itself and forms a *syrup* when concentrated. *Corn syrup* is partly degraded starch (83) consisting of individual glucose molecules as well as short chains of glucose molecules still joined together. It is formed by the action of enzymes in the bacterium *Aspergillus oryzae*, which breaks down the starch molecules; its viscosity is due to the entangling of the chains and their attraction to the surrounding water molecules. (*A. oryzae* is also used to prepare rice starches for fermentation into ethanol in the production of *sake*.) The syrup-forming ability of glucose is used in the manufacture of hard candy—a flavored and colored solution held together as a glassy solid by water-attracting glucose (and sucrose) molecules that do not allow the little water that remains to drip away.

. . . .

FRUCTOSE (80) $C_6H_{12}O_6$

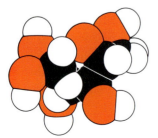

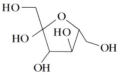

Fructose is another sugar that, like glucose, consists of a single chain of six carbon atoms. In its open-chain form it differs from glucose in the location of the doubly bonded oxygen atom, which is attached to the next to last carbon atom of its chain. Like glucose, it also forms six- and five-member rings (the latter is shown in the illustration).

Fructose is also known as *levulose* and *fruit sugar*. The former name comes from its effect on *polarized light: It behaves oppositely to glucose (dextrose), rotating the plane of polarization to the left. The name "fruit sugar" comes from its widespread occurrence in fruits and vegetables. Fructose is the major sugar in many forms of *honey*, for nectar (the fluid exuded by plants, usually flowers, perhaps to control the *osmotic pressure of their fluids and also to attract insects) contains a high proportion of the sugar.

Two features of fructose account for its properties and uses: It is about 50 percent sweeter than sucrose (81) and more soluble than both sucrose and glucose. The first property makes it useful in low-calorie diets, for with it the same amount of sweetening can be achieved with a smaller mass of carbohydrate. It may also, depending on its price, be more economical to use than sucrose. That fructose is more soluble than sucrose may account for the conversion of any sucrose in nectar to fructose, for then higher concentrations of carbohydrate may be achieved in a given mass of water. This greater solubility accounts for the softness of *brown sugar* as compared with pure, white sucrose; brown sugar consists of sucrose crystals coated with the glucose and fructose that remain in molasses after most sucrose has been removed by repeated crystallization. It also provides a neat way of making soft-centered chocolates: semisolid sucrose filling can be injected into the hollow center with enzymes (obtained from yeast) that convert sucrose to glucose and fructose; as the latter is formed, it dissolves in the little water remaining, and gives a soft, creamy texture to the filling. Since fructose retains water better than sucrose, it is used in jams and candies to reduce the chance of crystallization.

Commercial fructose is prepared from the glucose in corn syrup by enzymatic action (by *Streptomyces*), which rearranges the atoms of the glucose molecule into the marginally different fructose molecule. As well as providing a sweetener that can be used in smaller concentrations, this process has the additional economic advantage of using a readily available raw material.

Fructose is also the sugar that powers the motion of sperm. Men synthesize it in their seminal vesicles, and it is incorporated into semen and used by the sperm as fuel for their brief but portentous journey.

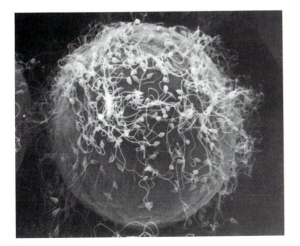

Sperm, such as these surrounding the egg of a sea urchin, derive their energy from fructose in the semen.

SUCROSE (81) C₁₂H₂₂O₁₁

SUCROSE (81) $C_{12}H_{22}O_{11}$

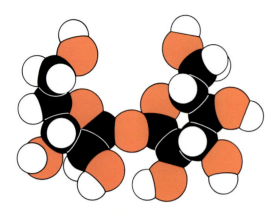

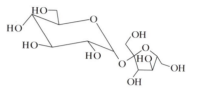

Sucrose is an example of a *disaccharide*, a molecule consisting of two glucoselike units (which are themselves *monosaccharides*) linked together. In sucrose, one of the units is a six-member glucose ring (79b), and the other is a five-member fructose ring (80). The two parts are linked by a chemical bond (so that sucrose is not a mixture, but a definite compound) at the points where glucose and fructose themselves would break open to give their open-chain forms. The sucrose molecule is therefore pinned together into a double-ring form, and it does not pop open in solution.

Sucrose is our common table sugar and one of the purest of everyday compounds. It occurs in most plant materials but is particularly abundant in sugar cane (hybrids of *Saccharum officinarum*) and sugar beet (*Beta vulgaris*), from which it is extracted by slicing, leaching, and refining. Beets provide the additional advantage to the farmer of having such deep roots that they mine nourishment from well below the surface, causing a circulation of nutrients. *Maple syrup*, the concentrated sap of the maple tree (*Acer*, especially the sugar maple, *A. saccharum*), is a solution of sugars, about 65 percent being sucrose, with small amounts of glucose and fructose. The sap is gathered from holes bored in the bark of the tree in early spring, pushed out when carbon dioxide, produced by metabolic activity and dissolved in water in the trunk, comes out of solution as the tree

A stem of sugar cane (*Saccharum officinarum*), the principal source of sucrose.

grows warmer in spring sunlight. (Most gases are less soluble in warm water than in cold, for the water molecules effectively shake the gas molecules out of solution as they begin to move more rapidly.) The brown color of maple syrup is the result of a reaction between the various sugars and the amino acids (73) that are also present.

When the enzyme *invertase* is added to a sucrose solution, the two parts of the sugar molecule are snipped apart, giving a solution of equal parts of glucose and fructose known as *invert sugar*. (The name reflects the effect of the sugar solution on *polarized light: Sucrose rotates the light in one direction and invert sugar in the other.) This mixture is sweeter than the parent sucrose because of the fructose.

When sugar is heated, a complex series of decompositions occurs. Each molecule already has many oxygen atoms (so that in a sense it is already partly burned), and their rearrangement is encouraged by the heat. The molecules break up, and the smaller fragments [including acrolein (102)] either vaporize immediately or dissolve and remain trapped in the complex solid residue called *caramel*, adding to its flavor.

. . . .

RAFFINOSE (82) $C_{18}H_{32}O_{16}$

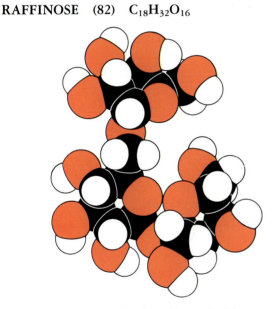

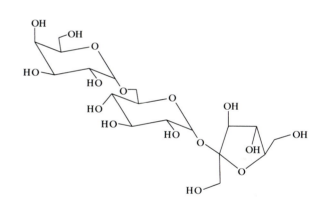

Raffinose is an example of an *oligosaccharide*, a sugar consisting of several glucoselike rings. The raffinose molecule is made up of a fructose ring (on the right in the illustration), a glucose ring, and the six-member ring of a sugar called *galactose* (which also occurs in milk).

When sucrose is ingested, it is broken down into its component sugars by *enzymes in the body. However, raffinose, which is found in peas and beans, is too much for our enzymes, and it passes undigested into the large intestine. There it is pounced on by hungry intestinal flora (including the bacterium *Escherichia coli*). They break it down and, in the process, release large amounts of gas—typically hydrogen, carbon dioxide (4), and methane (16)—in proportions that depend upon the person. Thus, meals of beans can result in considerable social inconvenience.

A similar failure to digest even disaccharides occurs in many individuals who, for genetic reasons, lack the enzyme *lactase*, which breaks down the disaccharide *lactose* that occurs in abundance in milk. In fact, production of

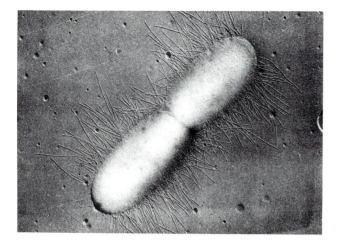

Our intestines swarm with *Escherichia coli,* one of which is shown here. They break down foods that we cannot digest and synthesize amino acids and vitamins, particularly vitamin K, which is necessary for blood clotting. If they invade the blood, however, they can induce disease, urinary tract infections, and septicemia.

this enzyme in adults appears to be the exception rather than the rule and is confined primarily to descendants of northern Europeans. It may have arisen in connection with calcium uptake and the whitening of the skin, which led to increased vitamin D production in the sunlight-poor north. (Vitamin D, or cholecalciferol, is a precursor of a substance that transports calcium and phosphate *ions through cell membranes and helps to build bones: A deficiency leads to *rickets*.) "Abnormal people," that is, people of northern European stock, can comfortably consume milk even after they are weaned; the "normal people" of Africa and the orient simply feed their colonic bacteria and suffer the pains of indigestion, which are often the pains of gas pressure in the colon. Almost no one has difficulty digesting cheese and yogurt because bacterial action eliminates much of their lactose during fermentation.

AMYLOSE (83)

AMYLOPECTIN (84)

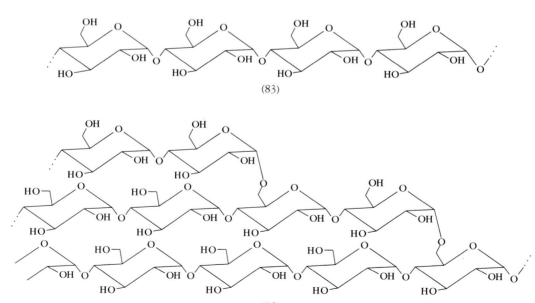

(83)

(84)

Now you will see the magic that nature conjures from the subtle deployment of meager resources—how, in a profound example of elegant economy, major components of organisms can be spun from the almost endless repetition of glucose (79) rings, and how a deft, genetically controlled twist of a bond can transform food to structure. You will see how a single molecular mutation (and hence its endless repetition, for that is the nature of life) can represent opportunity.

Starch is an example of a *polysaccharide*, a molecule consisting of many glucoselike units linked together into a polymer. It is a digestible polymer that is consumed in huge amounts worldwide in cereal grains and potatoes. It is present in such abundance in seeds (accounting for 75 percent of the weight of wheat flour) because the developing plant embryo requires a compact supply of energy. The first step in the digestion of starch is to cut the polymer chains into individual glucose molecules, which are then oxidized in cells and used to power growth, action, and thought.

Starch consists of two varieties of glucose polymer, *amylose* and *amylopectin*; the latter is the major component in most plants, comprising about three-fourths of the total starch in wheat flour. Amylose consists of long chains of glucose monomers, with very little branching. Amylopectin, on the other hand, consists of amyloselike chains with, occasionally, a different linkage to a glucose unit that results in a branch. This gives rise to a treelike structure, as shown in the illustration. Amylopectin molecules normally contain many more glucose units than amylose molecules. Both are nature's solution to the need for glucose for energy and the problem that glucose itself is so soluble. The solution is to link numerous glucose units together chemically, so that they are anchored to each other, yet in such a way that they can be snipped off by enzymes when a metabolic requirement arises.

Hydrogen bonding is part of the reason why starches are used in cooking. In natural starch, the molecules are bound tightly to each other by *hydrogen bonds, and they form compact solids that are encased in a membrane which is almost impenetrable to water. But when starch is heated in water, the water penetrates the granules and the hydrogen bonds between the molecules break down at about $65°C$; then water molecules flood into the solid and adhere all over the starch molecules by forming their own hydrogen bonds with the innumerable —OH groups. The starch molecules suddenly swell as the water molecules penetrate. In addition, their entanglement with each other and the adhesion of the water molecules together result in a sudden surge of viscosity. That is, the starch molecules thicken watery solutions, which can include gravy and sauces. However, on standing, the amylose chains exude and liberate water in a process called *retrogradation*; as they do so, they partially crystallize by sticking to each other again.

Cereal starches are essential to breadmaking. When dough is kneaded, the starch granules of the flour are broken down, and the enzymes that are present in the granules cut some of the polymers into sugar molecules. These are fermented by the added yeast (which is usually brewer's yeast, *Saccharomyces cerevisiae*), which converts the sugar molecules to alcohols and carbon dioxide, the former flavoring and the latter leavening the dough. The staling of bread is a form of retrogradation. Bread crumbs become hardened and bread becomes stale as the amylose chains and the linear amyloselike branches of amylopectin trees align with each other and crystallize.

Animals do not store their glucose as starch, but some is stored in muscle and in the liver as *glycogen*, which is closely related to amylopectin. The anaerobic metabolism of these reserves comes into play when the oxygen supply is reduced, as in sport, or terminated, as in death; in each case the metabolic product is lactic acid (33). In athletes, the resulting increase in acidity leads to cramp. In death, it leads to termination of the enzyme action that metabolizes the glycogen, consequent cessation of the transport of calcium ions, and hence the more permanent cramp of *rigor mortis*—the locking of muscles in their current state of extension or contraction. Glycogen levels are depleted less if animals are tranquil before they are slaughtered. The lactic acid levels are then higher and the meat is protected during storage by the mild preservative action of the acid.

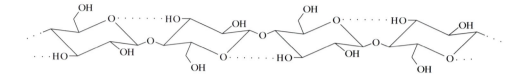

Although the monomers of cellulose are the same as those of starch, neighboring glucose units are linked differently. This results in long, flat, ribbonlike chains that are supported by *hydrogen bonds between neighboring units. These flat ribbons pack together, and the hydrogen bonds between them stabilize the structure into a solid, rigid mass. The difference between starch the fuel and cellulose the scaffolding—a simple twist of a link—shows unconscious nature at its most brilliant.

Cellulose, like starch, is a polysaccharide, but it is not digestible by humans. Ruminants need several stomachs

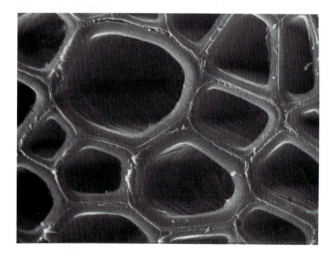

Cross section of the fibers of aspen, a hardwood (magnified 1760 times). These are typical springwood fibers with thin cellulose walls and large lumens. The boundaries between the fiber walls are lignin, which acts as a cement.

to digest grass, and even then they must draw on the cooperation of specialized fungi and other microorganisms in their gut, which cut up the chains and convert the glucose units to butanoic acid (39) and various other compounds. Although rabbits have only one stomach, they have developed the anatomically economical, but to us socially unacceptable, solution of multipass digestion, in which they eat some of their own excrement.

Natural cellulose is always found embedded in a matrix of *lignocellulose*, an amorphous mixture of *hemicellulose* (a polysaccharide formed from many different sugar molecules, with extensive branching into side chains) and *lignin* (a cross-linked polymer of *aromatic molecules). In papermaking and the production of cellulose fibers, the lignocellulose must be removed. Nature has done this almost completely in cotton, which has only a thin covering of lignocellulose. Wood, however, is heavily embedded, and wood pulp must be treated thoroughly. The lignocellulose is also responsible for the dark color of paper pulp, which must therefore be bleached before it can be used. The cheapest method, which is used for the pulp destined for newsprint, involves the *reduction (the opposite of oxidation) of the compounds responsible for the color. That is why newsprint turns yellow soon after it is exposed to air and light: Atmospheric oxygen reoxidizes the material, thereby undoing the bleaching accomplished by the reduction.

Purified wood cellulose can be converted into fibers if the majority of the hydrogen bonds between the chains can be eliminated. One approach is to convert many of the —OH groups to acetate groups (—O—CO—CH₃) by esterification with acetic acid (32). This removes the crucial hydrogen atoms, eliminates hydrogen bonding, and hence insulates each chain from its neighbors. The product is *cellulose acetate*, one of the cellulose derivatives used for textiles and photographic film. *Cellophane* is a cellulose sheet that has been purified and reconstituted. When you press on a cellophane sheet or try to tear it, you are experiencing yet again the strength of hydrogen bonds.

One of the few animals to produce cellulose—their outer wall is made up of *tunicin*, a form of cellulose—the tunicates (sea squirts, *Ciona intestinalis*) squirt nitric acid at an attacker as a defense strategy.

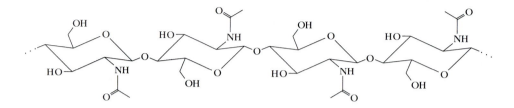

Nature's economical and elegant deftness with limited resources is shown in *chitin*, the structural material of the flexible inner parts of the exoskeletons of arthropods, including scorpions and crabs. Chitin also occurs in the beautiful, colorful, thin sheets we see as insect wings. It is identical to cellulose, except for the replacement of an —OH group on each glucose unit by an —NH(CO)CH₃ group. The nitrogen and sugar of chitin are valuable commodities, and some newly emerged insects postpone the taxing process of hunting by eating their discarded exoskeletons.

The cell walls of fungi are also of chitin, and not cellulose. The adaptation here may be that chitin is less liable to microbial degradation than cellulose. The advantage would stem from fungi having very high surface-to-volume ratios and hence being in almost total contact with their environment. This contact makes up for their immobility, for the fungus can scavenge for food by spreading through their surroundings. Some can do this enormously quickly, producing more than a kilometer of new mycelium (the collection of the individual filaments, or *hypae* that make up the fungus) in a day. The rigidity of chitin means that a fungus cannot engulf its prey but must attack it with *enzymes and absorb its components. These enzymes account for the decay wrought by fungi, but sometimes that decay is advantageous for it includes fermentation by yeast and the production of antibiotics.

Insect wings and the flexible inner parts of their skeletons are made of chitin.

Sauvignon blanc wine viewed in polarized light.

4

TASTE, SMELL, AND PAIN

This is the first of two chapters that deal with the chemistry of sensation. Here you will see how molecules can act as messengers between the external world and the internal universe of consciousness within our heads. All sensation is ultimately chemical, for all neuronal activity in our brains depends on the transport of molecules and ions from one location to another and the reactions in which they participate. But some messages from the outside world involve the direct perception of molecules that act as messengers. This direct perception of our surroundings is involved in taste, where the sensors are in the tongue; smell, where the sensors inhabit the nose; and certain varieties of pain, where the receptors pervade the skin and organs within the body.

In this chapter you will see that odor is closely related anatomically to emotion—that molecules can charm us into a recollection or a mood. You will also see many of the molecules that act as stimulators of taste, odor, and pain. These are the molecules to have in mind when you are sensing a flavor or a perfume, the molecules that, once you know them, may enhance the delight you find in a dinner or a person. These are the molecules of pleasure, warning, corruption, and communication.

SWEETNESS

Taste and smell, are examples of *chemoreception*. Taste in mammals is confined to the damp region inside the mouth, but some insects taste through their feet, and the bodies of fish are covered with chemoreceptors. Within the mouths of humans, chemoreceptors are largely confined to the mobile slab of muscular tissue called the tongue. Adult tongues are about 10 centimeters long and carry about 9000 *taste buds,* which are groups of 50 to 100 adapted *epithelial cells that are innervated by a smaller number of nerve endings. In this regard, taste is distinct from olfaction, for which the sensors are the actual nerve endings themselves (page 124). In the adult,

the taste buds are largely confined to the perimeter of the tongue, and their number declines with age, particularly after age 45. The tongues of children are covered by taste buds.

As shown in the illustration, different regions of the tongue respond to what some regard as the four basic tastes: sweet, salt, sour, and bitter. Sweetness, the taste considered in this section, is detected at the front of the tongue. Cats have very few "sweet receptors" and are among the few animals that do not prefer a sweet taste. Preference for sweetness and dislike of bitterness may be an evolutionary adaptation, since many ripe fruits are

The tongue is covered with papillae—this is an example of the variety called *fungiform*. They occur in the regions sensitive to sourness, saltiness, and sweetness. Taste buds are located on the top of the papillae, as at the center of the one shown here.

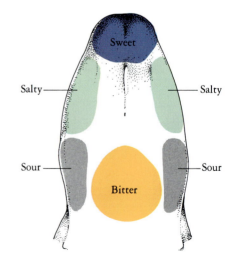

The regions of the tongue that respond to different components of taste.

TASTE, SMELL, AND PAIN

sweet (with ripeness, there is a decline of acidity, and the sweetness of sugars becomes more apparent) while many vegetable poisons are bitter.

Molecules that evoke taste are called *sapid* (from the Latin word *sapere* meaning "to taste"). One criterion of sapidity is solubility, because a substance must dissolve in water before it can penetrate into the taste buds. Particular tastes are evoked by molecules with groups of atoms in characteristic arrangements called *saporous units*. The saporous unit responsible for sweetness is called a *glucophore;* the structure of the glucophore presumably matches the structure of a protein in a taste receptor in a taste bud near the front of the tongue. When the molecule binds, perhaps by forming *hydrogen bonds with the protein, a signal is sent to the brain.

Several models have been proposed for glucophores. The problem is to identify a group of atoms in a particular geometrical arrangement which, if it is present in a molecule and the rest of the molecule is not too bulky to allow it to approach the receptor protein molecule closely, can bind to the protein molecule and result in the sensation of sweetness. The glucophore is thus a kind of molecular key, and the receptor protein a molecular lock. One model is shown in the illustration above. The red atoms labeled *A* and *B* must be electron-attracting atoms (usually oxygen, but sometimes nitrogen) that can participate in hydrogen bonding. Sweetness may also depend on the presence of a hydrocarbon group near the *A* and *B* atoms.

Fructose (80) is the sweetest sugar (when it is in six-member-ring form). Sucrose is generally experienced as being about one and a half times as sweet as glucose, which may be because it contains two glucophores in the right arrangement for them to fit two sites on neighbor-

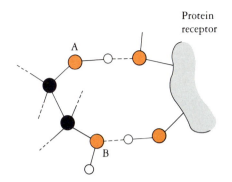

The presence of a unit of this shape with the distance between *B* and the hydrogen atom attached to *A* being 3×10^{-8} centimeters has been postulated as being responsible for the sensation of sweetness. It is the key that matches the protein lock of the appropriate sensor.

ing proteins. However, starch (83) is not sweet, even though it is rich in glucophores. This is probably because its individual glucose units are held back from fitting into the receptor sites by the rest of the chain.

Some small molecules, including glycols such as ethylene glycol (65) and glycerol (34), are sweet. The α-amino acids, such as glycine (73), are sweet, but amino acids with the —NH_2 group further removed from the carboxyl group are not: The neurotransmitter GABA (28) is tasteless. The tastes of mirror-image α-amino acids differ; D-amino acids (the ones that do not occur naturally) are all sweet, but the corresponding L-amino acids may be sweet, bitter, or tasteless. This emphasizes how important the shape of the molecule is in being able to attach to the receptor protein molecule: The mirror image of a molecular key might not fit the molecular lock. Some solutions of metal ions are sweet; for example, beryllium salts evoke a sweet taste, as do some lead salts (including lead acetate). Even solutions of common salt (sodium chloride) taste slightly sweet if they are extremely dilute.

SACCHARIN   (87) $C_7H_5O_3NS$

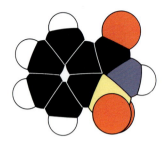

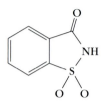

Saccharin was discovered in 1879 by a dirty, careless chemist who failed to wash his hands after a session in the laboratory. With its discovery came an opportunity for the food industry to satisfy the lust for sweetness without obesity, for saccharin is not metabolized by the body but is excreted unchanged. Saccharin went into commercial production in 1900. It is approximately 300 times sweeter than sugar but has a bitter, metallic after-taste that is difficult to mask. Its sweetness is lost if the hydrogen atom attached to the nitrogen atom is replaced by a —CH_3 group, which points to the involvement of the hydrogen atom in the saporous unit. Since saccharin itself is not very soluble, it is normally used in the form of the sodium or calcium *salt, either of which is more soluble than saccharin. Incidentally, saccharin does not fool bees or butterflies into treating it as sugar.

Saccharin is under suspicion as a carcinogen, since it has caused cancer in rat bladders. However, it is not metabolized by rats (or humans), whereas most carcinogens are. Moreover, since rats concentrate their urine more than humans do, it remains in their bladders for an unusually long time.

CYCLAMATE   (88) $C_6H_{12}O_3NS^-$

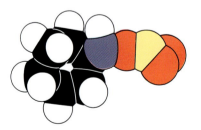

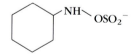

Cyclamate is used (in countries that still allow it) as the sodium *salt (or as the calcium salt in low-sodium diets) and, as such, is about 30 times sweeter than sucrose. It was discovered in 1937—like saccharin, by accident— in this case, by a careless chemist smoking a cigarette that had picked up some of the compound. Cyclamates were banned in 1969, perhaps unnecessarily, when it was found that massive doses of a mixture of drugs that included cyclamates led to bladder tumors in rats. This was traced to the formation of the known carcinogen *cyclohexylamine* as a result of the action of intestinal flora. Cyclohexylamine, a cyclohexane molecule composed of a ring of six —CH_2—groups with one hydrogen atom replaced by an —NH_2 group, is formed by loss of the —OSO_2 group from the cyclamate ion.

The cyclohexane ring provides the hydrocarbon group of the glucophore, and the sweetness is lost if the —NH— hydrogen atom is replaced or the cyclohexane ring is modified.

The water-repelling character of the cyclohexane ring is overcome by the ability of the —OSO_2 and —NH—groups to form *hydrogen bonds with water molecules, so the cyclamates are soluble in water.

....

ASPARTAME (89) $C_{14}H_{18}O_5N_2$

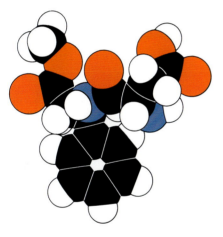

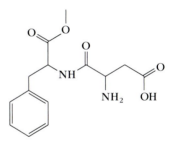

The aspartame molecule, a dipeptide, is a combination of two naturally occurring amino acids, aspartic acid and phenylalanine. It may be considered a tiny protein, because proteins are polypeptides (76). Aspartic acid is almost tasteless and phenylalanine is bitter, but their esterified dipeptide is quite different. It tastes 100 to 200 times sweeter than sucrose and lacks the unpleasant aftertaste of saccharin. However, as is typical of proteins, it is sensitive to heat and cannot be used in cooked foods. Aspartame also decomposes slowly in liquids, so that soft drinks sweetened with it have a limited shelf life. A mixture of saccharin and aspartame is sweeter and more stable than either substance on its own.

Aspartame is a white crystalline solid that was also discovered (in 1965) by accident—once again confirming that carelessness can be profitable as well as dangerous. In this case the careless chemist licked his dirty fingers and tasted sweetness. Because aspartame is a kind of protein, it is metabolized in the body like the other proteins we ingest and is a source of amino acids. However, since it is much sweeter than sucrose so that less has to be incorporated into food, it is less fattening than sucrose. Some people prefer its taste to that of sucrose.

SOURNESS AND BITTERNESS

Now we consider the sides and rear of the tongue, the sites where we taste sourness and bitterness.

Sourness is due to the presence of free hydrogen ions (H^+), which are released by acids such as the acetic acid (33) of vinegar, the phosphoric acid added to some cola drinks to enhance their zest, and the carbonic acid (4) of soda water. It is speculated that the taste buds on the sides of the tongue contain protein molecules rich in carboxylate groups ($-CO_2^-$), which are *carboxyl groups that have lost a hydrogen ion; in an acid medium, they are converted back to carboxyl groups and consequently cause a change in the shape of the protein molecules, which trigger impulses to the brain. It is quite easy for manufacturers to achieve the sensation of sourness simply by adding small concentrations of acid to their products.

Bitterness is often associated with the presence of organic compounds of nitrogen known as *alkaloids, which are widely present in the *angiosperms* (the flowering plants). Many of these alkaloids [which include strychnine, nicotine, and caffeine (152)] are poisonous, and the ability to detect them by taste may have arisen as an adaptation for survival. There is even speculation (it is no more than that) that the very limited ability of reptiles to detect bitterness may have contributed to the demise of the dinosaurs, which occurred at about the same time as the emergence of the angiosperms. That the detection of bitterness is an avoidance signal is supported by the observation that in only very few cases—quinine (92) and caffeine (152) among them—is bitterness sought for pleasure, and then only after training. The inclusion of some bitter principles in aperitifs may be a distant, highly domesticated echo of our ancestors' coping with

survival, for they stimulate the secretion of saliva. For us as modern sophisticates that is the prelude to a meal. For those who once needed to survive environments harsher even than cocktail parties, increased salivation may have been a last line of defense against poisons.

Perhaps because the food industry, always true to the echoes of our past, is more often concerned with the achievement of sweetness than bitterness, less is known about the receptors for bitterness than about those for sweetness. Some patterns have been discerned, including solubility (so as to pass through the saliva into the taste buds) and the presence of several $-NO_2$ groups in a molecule. One group of particular interest relates to the glucophore mentioned in the previous section. There, you saw that for sweetness the distance between the AH and B groups should be about 3×10^{-8} centimeter if the molecular key is to fit the sweetness lock. In some bitter compounds, the saporous unit is very similar to that of a glucophore but the AH-to-B distance is half that in the glucophore. If this is true, then we have an example of how a tiny change in a molecule can be the difference between pleasure and displeasure.

Some substances have keys that match both sweetness and bitterness receptors and can therefore evoke both tastes. One such molecule is found in the woody nightshade (*Solanum dulcamara*), which is also appropriately known as "bittersweet." Others include *ionic compounds and acids, where the cation may stimulate one response and the anion another. This is the case with salicylic acid (145): The hydrogen ion it releases in solution stimulates sourness receptors, but the accompanying salicylate anion stimulates a sense of sweetness that swamps the sourness; as a result, salicylic acid is sweet.

Solanum dulcamara is known as "bittersweet" (and also as "woody nightshade") because it elicits both components of taste. It is also poisonous.

. . . .

OXALIC ACID (90) $C_2H_2O_4$

CITRIC ACID (91) $C_6H_8O_7$

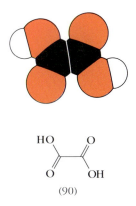

(90)

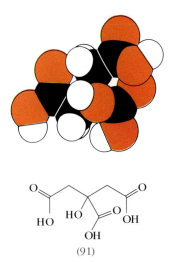

(91)

Oxalic acid occurs in appreciable concentrations in many leafy green plants, including rhubarb and spinach. The toxicity of rhubarb leaves was once ascribed to the presence of oxalic acid; however, spinach is rich in salts of the acid but is not toxic. It seems that the compound responsible for rhubarb-leaf toxicity has not yet been identified.

Apples (the fruit of trees of the genus *Malus*) are rich in *malic acid,* which is closely related to oxalic acid but has a —CH(OH)— group separating the two carboxylic acid groups; when you taste their sharpness, you should have this molecule in mind.

Citrus fruits are particularly rich in citric acid, lemon being the most concentrated, then grapefruit, and finally oranges. Citric acid is added to lemonade, and the discerning drinker can taste both the hydrogen ions it provides, which account for its sharpness, and the anions, which stimulate the sweetness receptors.

QUININE (92) $C_{20}H_{24}O_2N_2$

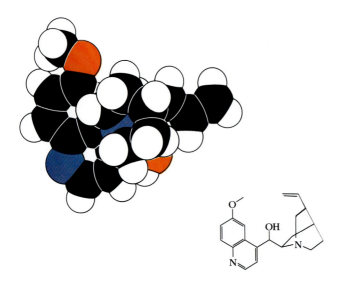

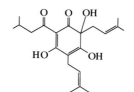

Quinine, a white, crystalline solid (like most *alkaloids) is extracted from the bark of the *cinchona* trees of South America. In countries unaffected by malaria, its bitter taste is most familiar in the form of the *tonic water* used in, among other things, gin and tonic drinks [and in which its taste is usually enhanced by a little citric acid (91)]. It is also a contributor to the taste of Dubonnet.

Quinine's more serious application, the abatement of malaria, depends on the ability of the quinine molecule to bind to DNA and hence to inhibit its replication. It affects only infected cells because they absorb the molecule in higher concentrations than unaffected cells. Quinine also has a mildly analgesic action. In doses stronger than that of a gin and tonic, it causes contractions of the uterine muscles and induces abortion.

HUMULONE (93) $C_{21}H_{30}O_5$

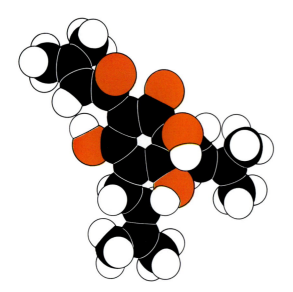

The yeast fermentation of cereal carbohydrates is the basis of the production of *beer*. In this process, starch molecules (83) from the kernels of barley are cut into individual monosaccharide and disaccharide (81) units by the action of enzymes that are naturally present in the sprouted grain. This is the process of *malting* (the term comes from an old word for "softening," as in melting), and occurs naturally to provide the sprouting seedling with an energy supply in the form of glucose. Much the same initial treatment is needed for the rice starch used in the production of *sake* (which is also technically a beer although commonly called a "rice wine"), but in this case the predigestion is due to the *Aspergillus oryzae* that grows on the rice. The germination is terminated by dehydrating the malt at a suitable stage. This "kilning" process also results in browning reactions (page 158) that contribute to the beer's final color, with the more heavily roasted malt resulting in darker beers, such as stout and porter. Some malts are bleached with sulfur dioxide to give a lighter beer. The product of this initial process is the *mash*, a concentrated liquid that is then separated into a solid part and a liquid part known as the *wort*. At this stage the hop resin is added to the wort. Hop resin is a viscous yellow substance contained in glands at the base of the blossoms of the female hop plant *(Humulus lupulus)* a relative of the cannibis plant (153); it is rich in bitter components, including humulone and its relative lupulone, which are extracted by the wort and which serve to offset the insipidity and sweetness of unhopped, sugary beer. The bitter molecules probably undergo a slight molecular rearrangement in the mixture in which the six-member ring breaks open and reforms as a five-member ring, producing *isomeric molecules that are even more bitter than humulone and lupulone themselves. Yeast is then added to the wort; it converts the sugars to ethanol (27) by fermentation, releasing carbon dioxide in the process. The technique of fermentation may be either "top fermentation" or "bottom fermentation." These techniques differ in the region where the used yeast cells accumulate. The former is used to produce English beers. These are more acidic and stronger than other beers partly because of the access that the yeast cells have to air. Bottom fermentation, the conventional procedure in the United States, is used to produce the lighter *lager*. The name (from the German word for "store") reflects the process used to clear the beer, that of storing it at just above the freezing point.

When beer is left exposed to sunlight, a *photochemical reaction occurs in which the humulone molecules react with other molecules that contain sulfur atoms, resulting in the formation of several sulfur-containing products. One of these is 3-methylbutane-1-thiol (129), a molecule manufactured even more liberally by the skunk.

Blossoms of the female hop plant (*Humulus lupulus*).

HOT, SPICY, AND COOL

Hot, spicy, and cool tastes are chemical stimulations of pain. There are two types of pain nerve: Class A nerves are slender fibers that carry signals rapidly (at about 20 meters per second); class C nerves are thicker and carry signals more slowly (at about 1 meter per second). Their signals are referred to as *fast pain* and *slow pain,* respectively. Fast pain is the response to injury and is often sharply localized. Slow pain is often a dull, aching sensation that is usually less sharply localized.

Both types of nerve fiber enter the spinal cord, along with nerves responsible for temperature sensation; there, they stimulate neurons that lead to the brain and their signals undergo some local processing. An important feature of pain nerves is the interaction of the two types in a gelatinous part of the spinal cord called the *substantia gelatinosa.* Signals arriving along the A fibers excite cells of the *substantia gelatinosa,* but those arriving along the C fibers inhibit them. The net effect can be to inhibit the cells that are responsible for transmitting A and C signals to their processing center in the brain (the thalamus). Hence there is a complex interplay between the signals arriving initially as fast and slow pain (a point

that will be illustrated in what follows). Moreover, in response to pain signals, the brain can secrete its own analgesics—the *endorphins* and *enkephalins.* Both are polypeptides (76) (with the endorphins having long chains and the enkephalins short chains) that affect the transmission of nerve signals, and both are mimicked by opiates (149). The pain receptors that initiate all this complex signaling are highly branched nerve endings themselves: There are no specific innervated pain receptors.

There are, however, receptors that respond to thermal stimulation. They are essentially of two types, one of which responds to hot and the other to cold; the latter are more numerous by a factor of about ten. Their signals, like pain signals, are carried by class A and class C nerve fibers, ultimately to the thalamus, so that intense thermal stimulation can be interpreted as pain.

Many of the spices used in curries and other foods stimulate pain-detecting nerve endings in the mouth (and elsewhere), but the relation between molecular structure and response is not known.

PIPERINE (94) $C_{17}H_{19}O_3N$

CAPSAICIN (95) $C_{18}H_{27}O_3N$

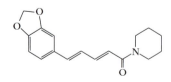

(94)

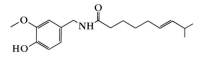

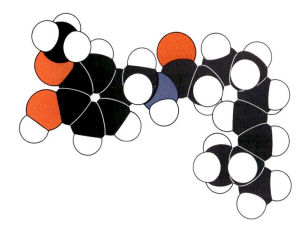

(95)

Piperine is the active component of white and black pepper (the berries of the tropical vine *Piper nigrum*). Black pepper is obtained by allowing the unripe fruit to ferment through the action of the fungus *Glomerella cingulata* and then drying it. White pepper is obtained by removing the skins and pulp of the ripe berries and drying the seeds. Piperine is an *alkaloid, and its presence presumably is an adaptation for the protection of the plant.

Another "hot" spice is capsaicin, the pungent component of various species of *Capsicum,* including red and green chili peppers and especially *C. annum* and the small, virulent *C. frutescens.* It is the active component of *paprika*.

The action of capsaicin (and perhaps piperine) seems to have several components. It stimulates the excretion of saliva, which aids digestion. It also stimulates movement of the colon and helps to encourage the passage of the remains of food. At the anus, it can cause itching (*pruritis*) and a sensation of warmth during defecation. The sense of well-being one gets after a meal of hot spices has been ascribed (in a speculation) to the ability of these pain-producing compounds to stimulate the release of soothing endorphins in the brain.

Red chilis (*Capsicum*) are pungent on account of the capsaicinoids they contain. One of the most abundant capsaicinoid molecules is capsaicin itself.

ZINGERONE (96) $C_{11}H_{14}O_3$

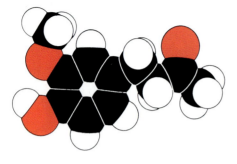

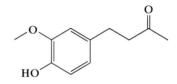

Although the relation between molecular structure and pain response is not known, it almost certainly has something to do with the shape of the molecule, which acts to fit a protein in the wall of the pain nerve ending: The key fits the lock, the protein changes shape, and a signal is on its way. Some confirmation of this idea comes from a comparison of the shapes of the capsaicin molecule (95) and the zingerone molecule shown here. Notice how closely they resemble each other: The zingerone molecule lacks a hydrocarbon tail and a nitrogen atom, but is otherwise very similar to capsaicin.

Zingerone is the pungent, hot component of *ginger,* the rhizome (or underground stem) of *Zingiber officinale.*

Closely related compounds are also present in ginger; they differ mainly in the length of the chain that replaces one of the hydrogen atoms of the terminal —CH_3 group (on the right in the illustration). Different forms of ginger have different proportions of these related compounds. When green root ginger is dried and powdered, it loses not merely its free water but also an H and an OH unit (as H_2O) from neighboring atoms of a side chain, leaving a double bond between the two atoms and a tail with a different shape. This affects the flavor slightly because it changes the composition of the zingeronelike mixture.

MENTHOL (97) $C_{10}H_{20}O$

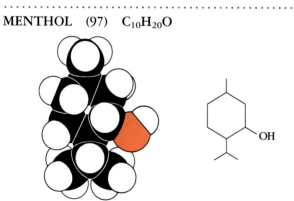

Menthol, from *oil of mint,* has a characteristically cool taste. It is extracted from the Japanese peppermint *(Mentha arvensis)* by cutting the plant when it blooms, curing

it like hay, and then distilling the oil in steam. Menthol is also prepared synthetically from turpentine (125). It is present in the common mint herb *(Mentha piperita)* and is used in cigarettes, soaps, and perfumes for the pleasant odor that accompanies its cooling taste.

The cooling taste arises from menthol's effect on cold temperature receptors in the skin. When menthol is present, the sensors responsible for signaling "cold" become active at a higher temperature than normal. Hence, an environment that is actually warm (such as the mouth) may be interpreted as being cool.

TASTE, SMELL, AND PAIN

MEATINESS AND BARBECUES

Most meat is muscle tissue and consists largely of proteins, so its principal components are strings of amino acids (73). Hence many of the properties of meat are the properties of its amino acids. The workhorse proteins of muscle are *actin* and *myosin,* which lie in layers between each other and slide past each other when the muscle is stimulated to contract. Contraction is maintained by temporary chemical bonds that form between the two proteins. The muscle fibrils that contain the actin and myosin molecules and the muscle fibers that are formed by groups of fibrils are encased in connective tissue, which is principally the protein *collagen.* A collagenlike protein is the principal structural component of commercial sponges (members of the class *Demospongiae* of the phylum *Porifera*), which makes them inedibly tough and gristly; fish muscles have very little collagen and are correspondingly very tender. When collagen is heated in boiling water, it forms *gelatin.*

Muscle contraction must be fueled as well as stimulated. The energy resource used depends on the muscle type and affects the appearance of meat, especially its color. There are two broad classes of muscle fiber: fast and slow. Fast muscle fiber (which is also called *white fiber*) is used for rapid motion and uses as its fuel carbohydrates, particularly glucose (79). This is readily available in the blood and as glycogen and can be used (albeit inefficiently) even without oxygen. Slow muscle fiber (which is also called *red fiber*) is used for steadier sustained motion. Its energy supply is stored as fat (36), for which oxygen is essential. Hence the oxygen supplied by the blood needs to be stored in slow muscle fiber, so that it can continue to function at least briefly even when the oxygen supply is insufficient to meet the demands of sustained action. The storage molecule is another protein, *myoglobin,* which is closely related to the oxygen carrier hemoglobin (page 90) and contains an iron atom

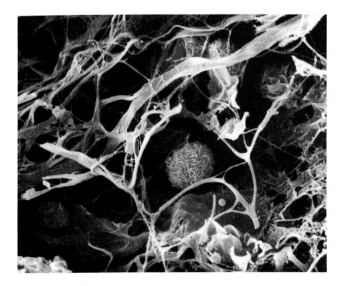

Connective tissue in the body typically consists of cells called *fibroblasts* that excrete a molecule, tropocollagen, that *polymerizes into collagen, the most abundant protein in the body. The irregular network of fibers shown here is collagen with embedded fibroblasts.

at its heart. Myoglobin is red when oxygenated, which is why slow muscle fiber is red while fast muscle fiber, which does not need this storage molecule, is pale.

Fish (especially cod, which spend most of the time lying around on the sea floor) have skeletal muscles that are built predominantly of fast muscle fiber. Hence they have pale, even white, flesh. Poultry, which are largely earthbound, have unexercised wings and hence white breast meat but pinker legs; the legs are greasier than breast meat because fat is the fuel of their red muscle fiber. Game birds fly frequently, require pectoral muscles that can sustain prolonged activity, and hence are myoglobined throughout. The muscle of most domesticated animals is slow and hence red. Diving animals, especially whales, require extensive oxygen reserves, and their flesh is very dark.

When myoglobin loses its oxygen after death, it becomes pale purple. When cooked it turns brown, as the iron at its center is *oxidized and, in its new form, absorbs light of a different *wavelength. When meat is salted, as in the preparation of ham and some cooked meats, the myoglobin molecule picks up nitrite ions (NO_2^-), and its color changes to pink.

The different tastes and odors of animal muscles are modified by cooking, which breaks proteins and other components into smaller fragments. Some of these fragments are small enough to be volatile and thus add to our perception of the flavor. Some originate in molecules of the substance called ATP, with which we begin our description.

White and red meat, as in this fish and beef, differ in the proportions of fast and slow muscle fiber.

TASTE, SMELL, AND PAIN

ADENOSINE TRIPHOSPHATE   (98) $C_{10}H_{11}O_{13}N_5P_3$

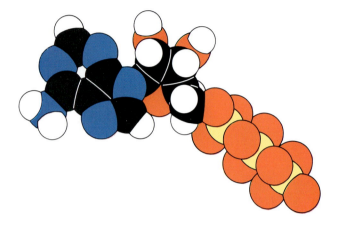

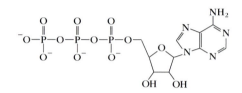

The adenosine triphosphate (ATP) molecule looks complicated, but it can be pictured as being formed from three components. The central component is a sugar molecule, *ribose,* in the form of a five-member ring resembling one form of the fructose molecule (80). Attached to this are the two other components. One is a group consisting of linked five- and six-member rings of carbon and nitrogen atoms. Such groups are called *bases,* and this particular one is *adenine.* The combination of the ribose and the base make up a unit called a *nucleoside.* Another atom of the ribose ring is attached to a string of three phosphate groups. As far as we are concerned here, this string is where the action is.

ATP is one of the most important molecules in living things, and it is abundant in muscles. Its role is as an immediate source of energy for powering biochemical reactions, such as may be needed for the contraction of muscle cells when lifting a load. It is also used much more broadly: for example, in the metabolism of foods, in the construction of proteins from the DNA template,

and in powering the processes that allow you to see this word and reflect on its significance. Indeed, where there is life, there is ATP. Its ubiquity explains why phosphates are so important in the diets of plants and animals. The phosphate fertilizer industry, in essence, converts old bones to an assimilable form of phosphorus that can be used to maintain the presence of ATP in growing, living plant cells.

ATP performs its role by releasing the terminal phosphate group at the demand of an enzyme; in so doing it releases energy to drive some other reaction, which may be the construction of a protein (76) from amino acids (73), or the contraction of a muscle. The liberated phosphate group is then reattached to the molecule using energy derived from ingested food. After death, the ATP is no longer reconstructed, and the muscles stiffen into *rigor mortis* (page 101). Rigor sets in more quickly if the ATP supplies have been depleted in a fight, a struggle, or even a state of anxiety before death.

INOSINE MONOPHOSPHATE   (99) $C_{10}H_{11}O_8N_4P$

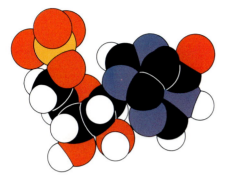

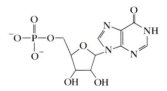

After death, when the reconstruction of ATP ceases, its decomposition goes beyond the loss of a single phosphate group. After losing two phosphates from its tail, to form adenosine monophosphate, it loses one of its nitrogen atoms, which is replaced by an oxygen atom. This produces a molecule of inosine monophosphate (IMP), a substance with a slight meaty flavor.

MONOSODIUM GLUTAMATE   (100) $C_5H_8O_4NNa$

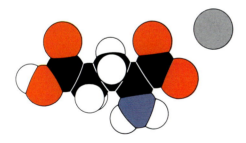

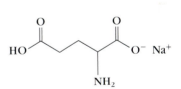

The more common sodium salt of this naturally occurring amino acid is *monosodium glutamate,* or *MSG*. It occurs in meat as the meat ages and its proteins decompose. Like IMP (99), MSG does not have a very pronounced meaty flavor on its own; but the two together are strongly meaty and are the principal compounds responsible for meat's taste. MSG is the cheaper and more readily available of the two; it is added by producers of meat foods to bring out the flavor of their product. It appears to enhance the sensitivity of the salt and bitter taste receptors of the tongue (page 110), but its precise mode of action remains obscure.

Different meats contain different proportions of IMP and MSG, and their flavors differ accordingly. Beef has

These oyster mushrooms, *Pleurotus ostreatus*, have a slightly meaty flavor partly because they share with meat proteins that are rich in glutamic acid.

twice as much MSG as pork, but approximately the same amount of IMP. Mushrooms are also rich in proteins composed of glutamic acid, which accounts for their slightly meaty flavor and their ability to enhance the flavor of most dishes. The part of the mushroom we eat is the *basidiocarp*, the fleshy, spore-producing body.

Although most microorganisms conserve their precious nitrogen atoms, there are several strains of bacteria (*Micrococcus* and *Brevibacterium* among them) that excrete glutamic acid when fed a diet rich in ammonia (7). This "fermentation" of ammonia is now the standard commercial method for producing MSG.

HYDROGEN SULFIDE   (101)   H$_2$S

The hydrogen sulfide molecule is the sulfur analogue of water (6), with a central sulfur atom in place of the central oxygen atom. A number of striking differences arise from this simple replacement.

To begin with, hydrogen sulfide is a foul-smelling poisonous gas. It is a gas at normal temperatures because sulfur is a less aggressive seeker after electrons than oxygen (it is a bigger atom, and the attracting nucleus is buried under clouds of electrons). Consequently, the nuclei of its hydrogen atoms are less exposed than in the water molecule and are much less able to take part in *hydrogen-bond formation. Hydrogen sulfide molecules therefore interact much less strongly with each other and are able to move freely as a gas.

Why hydrogen sulfide and many volatile sulfur compounds are foul-smelling is less easy to explain (see page 124). I should note here that one's perception of the gas declines after a brief exposure, which is very dangerous because hydrogen sulfide is more poisonous than hydrogen cyanide (104). In a sense this resembles a similar response to water: Since we are pervaded by water from conception to cremation, its smell is undetectable to us.

Hydrogen sulfide forms when sulfur-containing proteins decompose, either after death or during cooking. It has the characteristic smell of rotten eggs (and, in smaller quantities, the pleasant smell of a freshly boiled egg) and is indeed formed by eggs as their sulfur-rich albumin protein molecules decompose. The pale green coloration where the white of a boiled egg meets the yolk is another sign of its presence: The green is a precipitate of iron sulfide that forms where the hydrogen sulfide released from the albumin meets the iron-rich proteins of

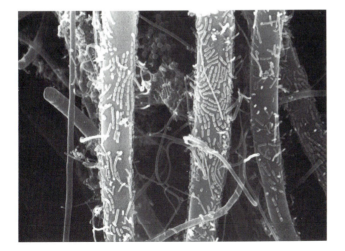

Bacteria clinging to *Beggiatoa* filaments growing near a warm ocean vent and living off the hydrogen sulfide in the water.

the yolk. Hydrogen sulfide reacts with many of the other compounds produced during the thermal degradation induced by cooking, and it gives rise to numerous other odorous molecules. It is particularly important in producing molecules that contribute to the flavors of cooked chicken.

Another fate for hydrogen sulfide is its conversion to elemental sulfur, just as water was the source of atmospheric oxygen. In much the same way as cyanobacteria learned to snip hydrogen atoms off water molecules and excrete the oxygen, which wafted off as a gas of O_2 molecules, so purple sulfur bacteria *(Rhodospirillum rubrum)* have learned how to snip hydrogen atoms off hydrogen sulfide molecules. What they excrete is solid elemental sulfur, which lies around in great piles and does not waft away. These bacterial dung hills are the deposits once quarried for sulfur.

· ·

ACROLEIN (102) C_3H_4O

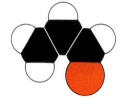

Acrolein, an aldehyde, is a colorless volatile liquid with an acrid smell. It is formed when the fatty acids (35) present in meat break down under the stress of heating: The long-chain fatty acids break off their glycerol anchor in the storm of thermal motion caused by the heat, and the glycerol molecule itself loses two molecules of water, forming acrolein. Acrolein's acrid smell adds considerably to what some regard as the pleasure of a barbecue, where it is easily detected in the smoke. It also contributes to the flavor of the caramel prepared by heating and partially decomposing sucrose (81).

Wood smoke also includes the simplest aldehyde of all—formaldehyde (29). Its attack on the eyes is partly responsible for the tears elicited by smoke. And its attack on bacteria is responsible for the preservation of smoked meats. Other substances in smoke include *phenols; they act as antioxidants (42), helping to protect fats from oxidation by diverting the attack of atmospheric oxygen.

FRUITS AND FOODS

The flavor of food is a combined response to two chemical senses, taste and odor. Taste has already been considered (page 106), and here we concentrate on odor. This is the more sensitive sense, even in humans, and it is the major contributor to the perception of flavor.

Whether or not a molecule has an odor depends on whether it can excite the olfactory nerve endings in the nose. In humans, these nerve endings occupy an area of yellow-brown *epithelium of about 5 square centimeters; except in sniffing, it is eddy currents (and not direct drafts) that carry molecules to this area.

Smell is peculiar: The 50 million or so receptors of the olfactory epithelium are bare nerve endings. This is unlike the case of any other sensation (except pain), where normally some kind of transducer acts as a buffer between the outside world and the nervous system. With smell, the nervous system is in direct contact with the outside world: In essence, the brain is exposed in the nose. This suggests that olfaction is one of the oldest and most primitive senses. Smell is also closely linked, at least in terms of the physical proximity of their processing centers, with one of the more primitive parts of the brain—the *limbic system*, the seat of the control of emotions. This may account for the powerful, and sometimes unconscious, impact of an odor.

Smelly molecules are called either *odorivectors* or *osmophores* (from the Greek word for "smell"). However, the relation between their molecular structure and the sensation they excite remains obscure. One criterion for smell is that an odorivector must be volatile, for otherwise it would never reach the nose. A second is that it should be at least slightly soluble in water, for otherwise

A portion of the surface of the olfactory epithelium at the top of the nose. Hairlike cilia originate from bodies located beneath the surface of the bulbous bodies, the olfactory vesicles.

it would not dissolve in the *mucus*, an aqueous solution of proteins and carbohydrates that is exuded by cells in the olfactory epithelium and coats the nerve endings there. However, this is a less clear-cut criterion, because organic molecules in the mucus could act as detergents (page 61) and carry insoluble molecules through the water and into receptor sites. An odorivector must also, presumably, interact with a protein molecule in the ol-

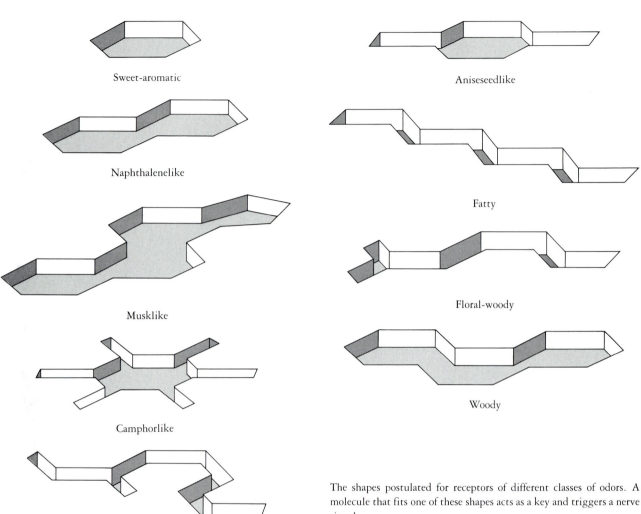

Sweet-aromatic

Naphthalenelike

Musklike

Camphorlike

Jasmine- and funguslike

Aniseedlike

Fatty

Floral-woody

Woody

The shapes postulated for receptors of different classes of odors. A molecule that fits one of these shapes acts as a key and triggers a nerve signal.

factory nerve endings, modify its shape, and hence stimulate the nerve cell to send a message to the brain.

It seems plausible that the interaction between an odorivector and a nerve-cell protein depends on the same kind of lock-and-key mechanism that activates taste: A molecule of a particular shape can attach to a given protein molecule so long as it matches its shape in some respects. The difficulty with the testing and develop-

ment of this idea is that there are so many different odors, and hence so many different protein locks. About 30 different types of anosmia—partial odor blindness—have been identified, which suggests that there are at least 30 different locks that can be opened. The shapes of a few of them have been identified (but not precisely) and are shown in the illustration above. Only a part of a molecule needs to fit snugly into a site to trigger a signal;

indeed, if the molecule is flexible, it may be able to fit into more than one site, thereby exciting a mixed response.

Pungency and putridity are less specific than other odors; they may be responses that are signaled by cells other than the ones described here. Lurking in the olfactory epithelium, among the mucus-exuding cells, are cells that are part of the system that innervates the face (the *trigeminal nerve*). It is suspected that pungent and putrid molecules penetrate into them, interact with their proteins, and stimulate them to fire. Thus there may be two types of olfaction: "first smell," the ordinary variety for specific odors, and "second smell," for nonspecific pungency and putridity.

The molecules discussed in this section are related to typical fruits and foods. The essential oils of plants and parts of bodies are also rich sources of olfactory stimulation, for good or for ill, and are described in the following two sections.

· ·

BENZALDEHYDE (103) C₇H₆O

$$BENZALDEHYDE \quad (103) \quad C_7H_6O$$

HYDROGEN CYANIDE (104) HCN

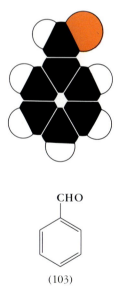

CHO

(103)

(104)

Benzaldehyde is a colorless liquid that smells of bitter almonds. A closely related molecule, phenylethanal, is obtained conceptually by inserting a —CH₂— group between the benzene ring and the —CHO group. This latter molecule fits a floral receptor better than benzaldehyde itself. It smells of hyacinth and is used in perfumes under the name *hyacinthin*.

Hydrogen cyanide is an almond-smelling, colorless, poisonous gas with an odor that fades on prolonged exposure. Hydrogen cyanide interferes with the transport of

oxygen by the red blood corpuscles and with the action of the energy-source molecule ATP (98), so that it brings the body to a stop. [The cyanide ion (CN⁻) has the same number of electrons as a molecule of carbon monoxide (CO), and so has similar chemical properties.]

The aroma of cherries and almonds is due to benzaldehyde, but the hydrogen cyanide in cherries, also contributes somewhat. Benzaldehyde and hydrogen cyanide both occur quite widely in drupes and pomes (multiple-seed and single-pit fruits), especially apricots and peaches. They are released when the pips are crushed and *enzymes can get to work. This much the Romans and Egyptians also knew, for they ground peach kernels to make poisons. The source of both molecules is the compound *amygdalin,* which is a modified disaccharide (specifically, a *glycoside*) consisting of two glucose (79) units linked together, with one of the —OH groups changed to a benzaldehydelike group with a —CN group attached. The enzyme *emulsin* can disrupt this molecule, releasing two glucose molecules, a benzaldehyde molecule, and a hydrogen cyanide molecule. Some cyanide appears in fruit jams that contain pit and pip extracts, such as quince. The cyanide is normally in too small a concentration to be dangerous.

ISOAMYL ACETATE (105) $C_7H_{14}O_2$

ETHYL 2-METHYLBUTANOATE (106) $C_7H_{14}O_2$

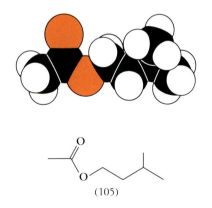

(105)

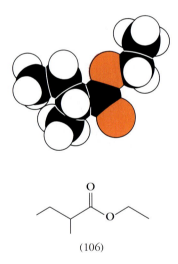

(106)

With these two molecules we see nature building different compounds in similar ways and from the same kit. The isoamyl acetate molecule is an *ester formed from acetic acid (32) and an alcohol, isoamyl alcohol. The ethyl 2-methylbutanoate molecule is also an ester with the same numbers of carbon, hydrogen, and oxygen atoms, but they are bonded in a different pattern.

Both compounds grow in prominence as apples ripen and, as their concentration increases, they mask the characteristic flavor of the unripe fruit. Esters with about seven carbon atoms have characteristic fruity smells, occur widely in fruits, and result from the breakdown of long-chain fatty acids (35) as the cell membranes are *oxidized during the ripening process.

2-HEPTANONE (107) $C_7H_{14}O$

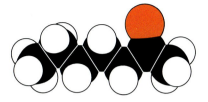

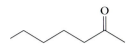

This compound is a liquid with a clovelike odor (it occurs in oil of clove). Its oxygen atom allows it to be transported through the mucus of the olfactory epithelium, and it fits neatly into the receptor characteristic of woody-fruity cells (and maybe others too, although I must stress that shape-odor relations are still little more than speculations). The presence of 2-heptanone accounts for the odors of many fruits and dairy products. The molecule is also responsible for the aroma of blue cheese, which is formed by inoculating the curing mixture with molds such as *Penicillium roquefortii*. Like butanedione (40), which gives butter, buttermilk, sour cream, and cottage cheese their characteristic odor, 2-heptanone is a ketone.

3-(*para*-HYDROXYPHENYL)-2-BUTANONE (108) $C_{10}H_{12}O_2$

IONONE (109) $C_{13}H_{20}O$

(108)

(109)

I include these two ketones here because they show how two molecules in a mixture can conspire to produce an aroma that may evoke memories of others. 3-(*para*-hydroxyphenyl)-2-butanone is the molecule chiefly responsible for the smell of ripe raspberries, and it is included in the recipe for the synthetic flavor. The fresh smell of the newly picked fruit is due partly to ionone, which is also responsible for the odors of sun-dried hay and violets.

Ionone is the fragrant component of oil of violets, which is obtained (by extraction in solvents) from the flowers of the blue and purple varieties of *Viola odorata*. Natural oil of violet is too expensive as a source, and most of the ionone molecules used in the perfume and food-additive industries are made synthetically.

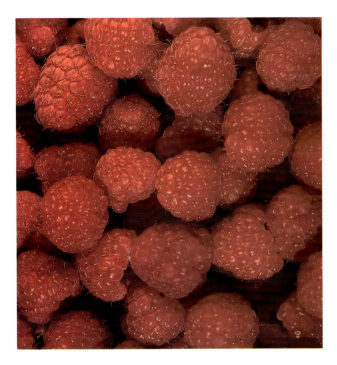

That fresh raspberries can evoke the smell of new-mown hay is no accident: they contain the same molecule (ionone) as hay.

METHYL 2-PYRIDYL KETONE   (110) C_7H_7ON

2-METHOXY-5-METHYLPYRAZINE   (111) $C_6H_8ON_2$

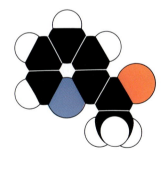

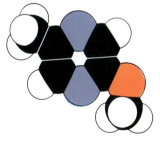

(110)

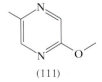

(111)

Molecules containing a benzenelike ring (but with one or more of the carbon atoms replaced by nitrogen atoms) play an important role in the aroma of heat-treated foods. The pyridyl molecule illustrated is responsible for the pervasive odor of popcorn. The pyrazine molecule is responsible for the odor of peanuts. Pyrazines also con-

tribute to the aromas of crusty bread, rum, whisky, chocolate, and some uncooked vegetables, including peppers. When you take delight from these aromas, you are responding to molecules like the ones shown here, as they stimulate the part of your brain that ends in your nose.

2-FURYLMETHANETHIOL   (112) C_5H_6OS

This molecule is one of those responsible for the aroma of *coffee,* the roasted beans of *Coffea arabica* (the plant first cultivated near Mocha in the Yemen) or of the more climatically tolerant *C.* *canephora.* (*Mocha* now means a mixture of coffee and chocolate.) The stimulating action of coffee is due to caffeine (152). See the description of that molecule for more details.

Roasted coffee beans. The color is largely due to the browning reaction (page 158) that occurs when organic substances containing nitrogen are heated. Temporarily trapped within the beans are the molecules responsible for flavor and stimulation.

DIALLYL DISULFIDE   (113) $C_6H_{10}S_2$

ALLYL PROPYL DISULFIDE   (114) $C_6H_{12}S_2$

THIOPROPIONALDEHYDE-*S*-OXIDE   (115) C_3H_6OS

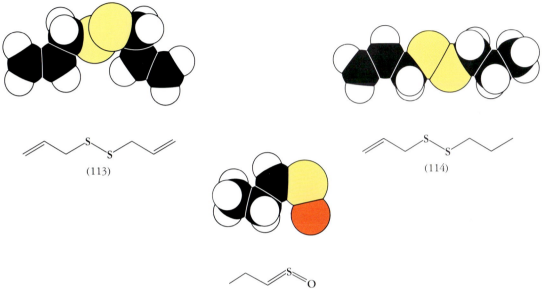

(113)

(114)

(115)

Sulfur compounds are responsible for the pungent odor of members of the genus *Allium,* including garlic *(A. sativum)* and onion *(A. cepa).* All these plants contain a high concentration of amino acids (73), particularly *cysteine,* that include sulfur atoms on the side chains. Garlic and onion are odorless until crushed and chopped; then the damage done to the cells allows enzymes to reach their contents and to convert nitrogen- and sulfur-containing amino acids to volatile compounds, including ammonia (7) and the compounds shown here. The smell of garlic powder is different from that of fresh garlic because it was acted on by enzymes in the past, and the most volatile compounds have had time to evaporate and to oxidize.

The aromas of species of *Allium* are due largely to disulfides, in which two sulfur atoms act as a bridge between two hydrocarbon fragments. Diallyl disulfide is responsible for the odor of garlic, and allyl propyl disul-

Garlic cloves (*A. sativum*). Pungent odors, including those of garlic and onion, are often due to molecules that contain sulfur atoms.

fide for that of the onion. Enzyme action on the onion amino acids also results in the formation of the thiopropionaldehyde-*S*-oxide, which is the lachrymatory component of chopped raw onions.

FLOWERS AND ESSENTIAL OILS

The *essential oil* of a plant is the essence of its fragrance. More prosaically, it is the volatile material that can be isolated from a single species of plant. It is often obtained by heating the leaves or petals in steam and is collected as the oily fraction. Some essential oils are pressed out of the plant, and others (particularly those from flowers) are extracted using solvents. About 3000 essential oils have been identified, and several hundred are available commercially. The subtlety of nature shows up here, however, for a natural essential oil may involve several hundred different types of molecules, and its impact may be changed by removing those in even very minute concentration.

A major use for essential oils is in the blending of perfumes, but they are also used to flavor foods. When next you wonder how a perfume can have so strong an impact, or bring back a memory, remember that olfactory signals are processed near the *limbic system, the part of the brain that is closely associated with emotion.

Most molecules that occur in essential oils contain up to about a dozen carbon atoms. This allows them to be moderately volatile, but to have a wide range of distinctive structures. A speculative connection between molecular shape and smell is outlined in the previous section. It may be possible to assess some essential oil molecules in light of the shapes of the olfactory receptors shown there; however, the possibility that a given molecule can evoke several responses by fitting into several receptors makes analysis of their odors almost impossible.

What follows is just a quick tour through the types of molecules encountered in different oils. Several of the molecules shown are members of the class of organic compounds called *terpenes*. These are compounds based on the isoprene unit (62), as illustrated; in a sense, they are very tiny, fragrant fragments of rubber (63), for rubber is a longer string of isoprene units. Terpenes are very common in plants, where they contribute not only to fragrance but also to color (page 150).

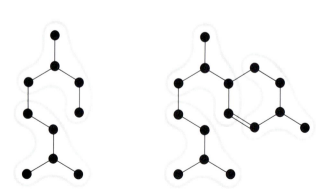

The common feature of terpenes is that they are constructed by stringing together 5-carbon-atom units based on isoprene.

BENZYL ACETATE　(116)　$C_9H_{10}O_2$

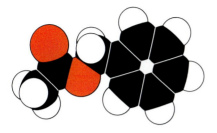

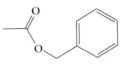

This molecule represents the class of essential oils that are *esters; benzyl acetate is an ester of acetic acid (32) and benzyl alcohol. The latter is a modification of methanol (26) in which a hydrogen atom from the —CH_3 group has been replaced by a benzene ring (23).

Benzyl acetate is one active component of *oil of jasmine;* another is *linalool.* Although jasmine (genus *Jasminum*) is a source of this compound, it is generally synthesized directly from acetic acid and benzyl alcohol. Since it is cheap and readily available, jasmine is a common perfume for toiletries.

CARVONE　(117)　$C_{10}H_{14}O$

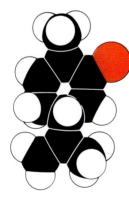

Carvone is representative of the terpenes. It is the main active component of *oil of spearmint* and is distilled from the leaves of the spearmint plant *(Mentha viridis),* a relative of common mint. Traditional chewing gum is flavored with spearmint. The gum part was originally *chicle,* the coagulated latex of the sapodilla tree *(Achras sapota),* but it is now made predominantly from a synthetic styrene-butadiene copolymer (64). The outer casing is dried sugar, held together by *hydrogen bonds until those bonds are replaced by water molecules from saliva.

A clever trick of molecular structure transforms spearmint flavor into caraway. Both molecules have the same atoms joined together into the same network. They differ in that one is the mirror image of the other, as a left hand is to a right (page 89). That a left-handed molecule should smell different from a right-handed version of the same molecule is consistent with the view that the olfactory receptors are sensitive to shape. One molecule is a key for particular receptor proteins; its mirror-image molecule may unlock different doors and hence different odors.

CINNAMALDEHYDE (118) C₉H₈O

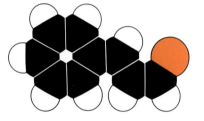

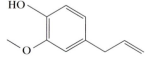

The cinnamaldehyde molecule is an example of a fragrant *aldehyde. It occurs in *oil of cinnamon* and is obtained by steam distillation of the bark of the cinnamon tree *(Cinnamomum zeylanicum)* and the leaves of the Chinese cinnamon. Cinnamon stick or powder is the dried inner bark. As well as for its fragrance, cinnamon has a reputation for its *carminative* action, its ability to release gases—hydrogen sulfide (101), methane (16), and hydrogen (H₂)—from the intestine and the stomach in one direction (belch) or the other, (politely, flatus), perhaps as a result of its irritant action.

EUGENOL (119) C₁₀H₁₂O₂

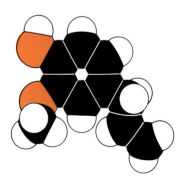

Eugenol occurs in *oil of bay* and is extracted from bay leaves—originally from the tall Mediterranean *Laurus nobilis* but now more commonly from the Californian *Umbellularia californica*. Eugenol is also an active component of *oil of clove*, which is obtained from the dried nail-like (*clavus* is Latin for nail) flower buds of *Eugenia aromatica*. It is, like cinnamon (118), a carminative compound.

Isoeugenol differs from eugenol only in the location of the double bond in the hydrocarbon side chain: This shifts the odor from clove to nutmeg (*Myristica fragrans*), one of the most important of the traditional spices. Nutmeg is derived from the ground seeds of the tree, and *mace* from the seed coatings.

A growing nutmeg in its red coat. The coat when dried and ground is mace.

GERANIOL (120) $C_{10}H_{18}O$

2-PHENYLETHANOL (121) $C_8H_{10}O$

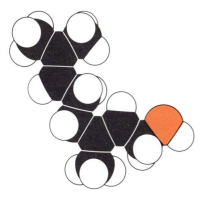

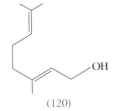

(120)

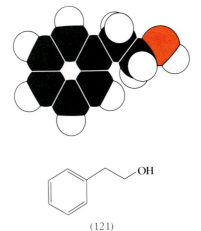

(121)

Geraniol occurs in many essential oils, such as the *citronella* obtained from Javanese citronella grass, and (as esters) in the leaves of the geranium. Its most delicate origin, though, is the rose, for together with 2-phenylethanol it is responsible for their fragrance and is extracted from their flowers. Now when you smell the fragrance of a rose, you will know what molecules are triggering a signal to your brain. But what goes on in your head, how your emotions respond, and what is the nature of enjoyment are still far from being understood.

....

ANETHOLE (122) C₁₀H₁₂O

ANETHOLE (122) $C_{10}H_{12}O$

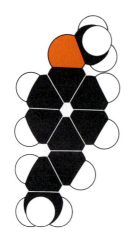

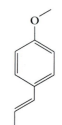

Anethole is the active component of *oil of aniseed (Pimpinella anisum)*. It also contributes to the flavor of fennel and tarragon. A close relative of anethole, anethole trithione, in which the hydrocarbon side chain is completed to form a ring with sulfur atoms, is one of perhaps a hundred substances that taste bitter to some Caucasian and Indian people and are tasteless to others; African, American Indian, Chinese, and Japanese people do not seem to display a taste blindness of this kind.

....

CAMPHOR (123) C₁₀H₁₆O

CAMPHOR (123) $C_{10}H_{16}O$

T A S T E ,   S M E L L ,   A N D   P A I N

Camphor, a white solid, is obtained by steam-distilling the trunk, roots, and branches of the camphor tree, *Cinnamomum camphora,* which grows in China and Japan. It is used as a counterirritant in medicine, like oil of wintergreen, and as an *antipruritic* (antiitch), perhaps because, like menthol (97), it selectively stimulates cold sensors. It was once used as an *analeptic* (from the Greek word for "restore"), a drug that stimulates the respiratory system, for it provokes deep breathing and raises the blood pressure. However, in larger doses camphor can lead to convulsions and respiratory collapse.

....

α-PINENE   (124) $C_{10}H_{16}$

α-TERPINEOL   (125) $C_{10}H_{18}O$

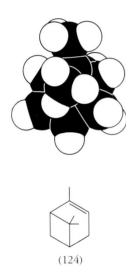

(124)

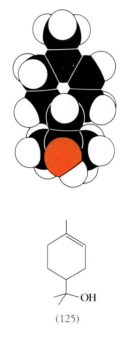

(125)

The α-pinene molecule is closely related to the camphor molecule (123), but it has no oxygen atom: α-pinene is a hydrocarbon. If you think of the awkward central "handle" of an α-pinene molecule bursting open, you have almost envisioned the α-terpineol molecule. It needs another —CH$_3$ group on the ring in place of a hydrogen atom, and an —OH group at one end of the broken handle.

Oil of turpentine is obtained by steam-distilling the resin exuded by various species of conifer trees, particularly (in the United States) longleaf pine (*Pinus palustris*) and slash pine (*P. elliottii*). Its major component is the fragrant hydrocarbon α-pinene. Although *pine oil* can be extracted naturally, it is often prepared by treating α-pinene with acid, which converts it to α-terpineol; the latter is the major component of pine oil and the molecule responsible for the fragrance of the juniper. Pine oil is the perfume and bactericide in many domestic cleaners.

VANILLIN   (126) $C_8H_8O_3$

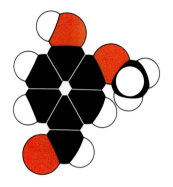

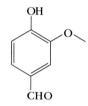

Vanillin is the essential component of *oil of vanilla,* which is extracted from the dried, fermented seed pods of the vanilla orchid *(Vanilla fragrans),* grown principally in Madagascar, Mexico, and Tahiti. During the curing process, the vanillin molecule is released from its *glycoside, a compound in which the molecule is attached to a molecule of sugar. Vanillin is one of the most widely used flavor compounds, and natural supplies are inadequate. It is therefore synthesized on a large scale; one method is by the oxidation of eugenol (119).

Vanillin can be detected in extremely low concentrations, yet the strength of its perception does not increase greatly as its concentration is raised (such is the oddness of olfaction). It is used in perfumery, in confectionary [chocolate (127) is a blend of vanillin and cacao], and for masking the smell of some manufactured goods. Vanillin molecules are leached out of the oak barrels (made from the European *Quercus robur* or *Q. sessilis,* not the American *Q. alba*) used to age wine, and contribute to the wine's "finish."

2,4-DIMETHYLPYRAZINE   (127) $C_6H_8N_2$

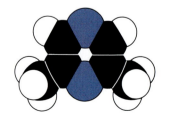

Such is the appetite for the taste of chocolate since its inception as *chocalatl,* the presumed aphrodisiacal drink of nobles in the Aztec court, that commerce has been stimulated to synthesize its flavor. One recipe depends on a mixture of vanillin (126) with an organic sulfide and 2,6-dimethylpyrazine, the molecule shown here.

Chocolate itself is obtained from the fermented seeds of *Theobroma cacao* (the name comes from the Greek words for "food of the gods"). After fermentation, they are roasted to remove volatile matter, including butanoic acid (39) and tannins (page 156), then ground, and finally mixed with sugar.

The sharp melting point of chocolate and its cool taste are due to its predominance of fats that have very similar hydrocarbon chain lengths, so that its softening resembles the melting of a pure solid rather than a mixture (page 55). Cocoa differs from chocolate in that it is made from cacao beans that have been pressed to remove some of the fats; the beverage made from it then does not acquire a fatty layer on top.

The main stimulant in chocolate is *theobromine,* which differs from caffeine (152) only in the replacement of one N—CH₃ group with N—H.

A pod of the cocoa plant (*Theobroma cacao*). Inside are the seeds that when fermented are the origin of chocolate.

ANIMAL SMELLS

Animals have a rich system of chemical signaling in which a substance called a *pheromone* is emitted by one and causes a response in another. Once again, the compulsion to respond may be due to the proximity of the region in the brain where olfactory sensations are processed and the limbic system, which is associated with emotion. Here chemistry has its most immediate impact on our emotions and behavior.

Many animals, including humans, continually emit a variety of molecules from their skin and sundry orifices. Indeed, over three hundred different compounds have been detected in human effluvia, one of the more abundant being, oddly, isoprene (62); but perhaps it is not so odd when we realize that isoprene units are part of the vitamin A molecules involved in vision.

The most obvious emanation from humans, other than their exhaled breath, is due to flatulence. People have to rid themselves of about half a liter of intestinal gas each day, about half of which is nitrogen (3) from the air, gulped down with food. Most of the rest is carbon dioxide (4), the product of the metabolism of organic matter in the intestine by the bacteria that inhabit it (among them, *Escherichia coli*), along with a little methane (16) and molecular hydrogen (H_2). So far, so good, for none of these gases smells. However, nitrogen atoms are abundant in proteins (page 88), and usually sulfur atoms as well. The fate of these atoms, once bacteria get to work on the amino acids (73–75) of these protein molecules, includes trace amounts of the pungent gas ammonia (7) and the foul-smelling gas hydrogen sulfide (101). They flavor the flatulence, and although one's own limbic system may respond favorably, that may not be true of one's neighbor's. The volume of gas may be enhanced if the food cannot be digested before it reaches the colon, as is the case with some oligosaccharides (82) found in beans. The odor may be enhanced if the food is rich in sulfur-containing amino acids and other sulfur compounds, as are varieties of the cabbage *(Brassica oleracea)*, especially Brussels sprouts.

The axillary region of the body (that is, the armpit) is a warm, moist region that is often a fount of odor. Once again bacteria are the cause, for the *Streptococcus albus* that inhabit the skin excrete lactic acid (33), thus increasing the acidity of their environment. This encourages other bacteria present to digest the organic components of sweat. The sweat itself is largely odorless, but the bacteria spice it with ammonia, hydrogen sulfide, and related compounds. Another component of male underarm sweat stimulates an engaging story. This component is a hormone molecule that closely resembles one secreted in the saliva of a male pig encouraging mating behavior in a sow. But that is not all. This same pheromone is also secreted by the fungus we know as the truffle, the fruiting body of the *Tuberales*. Because truffles do not appear above ground they must be sought out from their symbiotic cohabitation among the roots of certain trees, which pigs do, only to be frustrated, and which dogs and goats can be trained to do. Whether our enjoyment of truffles is related to our perhaps unconscious enjoyment of our own underarm sweat is a matter of conjecture.

CIVETONE (128) $C_{17}H_{30}O$

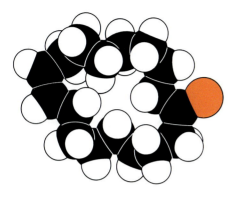

The civetone molecule does not lie in a perfect circle but can twist and turn (except at the carbon-carbon double bond) and adopt many different shapes. That perhaps enables it to slip into many different olfactory receptors. In particular, its 17-carbon-atom ring just slips into the musklike receptor (page 125).

Civetone is responsible for the sweet odor of *civet,* the soft, fatty secretion of the perineal gland of the civet cat *(Viverra civetta).* The civet, which is initially unpleasantly pungent, may be collected. The perfume glands are largest in the male and are located between the anus and the genitalia. Civetone has been used in perfumery for centuries, and most of it is now prepared synthetically.

The musk deer *(Moschus moschiferus)* of central Asia has a small sac in the skin of its abdomen in which a secretion collects as a viscous, brown, strong-smelling oil, especially at the time of rut. The secretion is complex and includes cholesterol (38), long-chain fatty acid esters (36), and a small amount of the ketone *muskone,* a ring-like molecule closely resembling civetone and largely responsible for its musky odor.

Musk is used in two ways in perfumery. It is used for its odor, as a component of heavy, musky, oriental perfumes. It is also used as a *fixative,* sometimes in concentrations so small that its own odor is masked. That is, it

is added to more volatile fragrances to retard their evaporation, so that they are experienced as a symphony of odors rather than as a sequence in order of decreasing volatility.

The civet cat *(Viverra civetta)* which inhabits Africa and Malaya. They are closely related to hyenas and are the source of civetone.

3-METHYLBUTANE-1-THIOL (129) C₅H₁₂S

3-METHYLBUTANE-1-THIOL (129) $C_5H_{12}S$

This molecule can be thought of as being like an alcohol (26), but with a sulfur atom in place of the oxygen atom. Such compounds are called *thiols* and were once called *mercaptans* (because they captured mercury atoms). This thiol bears out the reputation of sulfur compounds in general [as expressed in garlic (113) and onions (114)] for having fierce odors: 3-Methylbutane-1-thiol is the molecule squirted out in abundance by the striped skunk *(Mephitis mephitis)*. In the skunk, the anal sacs that contain the secretion flavored by this molecule are embedded in the muscle that erects the tail; thus, in a neatly eco-nomical piece of biological engineering, tail erection can be accompanied by an immediate squirt toward the adversary.

Sulfur compounds like the one shown, but with an —S—CH₃ group in place of —S—H, contribute to the odor of the urine of the red fox. Similarly, the odors of mink, stoat, and ferret are produced by cyclic compounds in which the hydrocarbon chain has bent around so that the carbon atom at its other end has replaced the hydrogen atom of the —S—H group.

These skunks, dead and labeled for a museum, once generated 3-methylbutane-1-thiol in abundance.

UREA   (130)   CH₄ON₂

Urea is the major organic component of human urine, the end product of the breakdown of the strings of amino acids (73) that constitute proteins (76). An adult excretes about 25 grams of urea each day. As urine goes stale, microbial action converts it to ammonia (7), with its typically pungent smell.

The cloudiness of the urine of cattle, horses, and some vegetarians is due to the precipitation of calcium and magnesium phosphates in the more alkaline solution that their diets produce. The golden yellow of urine is due to several molecules (specifically *uroporphyrin* and *stercobilin*) of the type shown in the illustrations below. Urine is said to *fluoresce in ultraviolet light—which is perhaps an as yet unexploited source of entertainment in men's rooms. The African bird *Turacos,* which is remarkable for its plumage, uses a version of uroporphyrin in which a copper atom lies between the four nitrogen atoms and provides the red color of its feathers.

The color of feces is largely due to stercobilin, a molecule that results from a long series of changes: First is the decomposition of some blood, then its conversion in the liver to the pigment *bilirubin* of bile, then the conversion of bilirubin into a colorless compound, and finally the *oxidation of that compound to give the yellow-orange stercobilin.

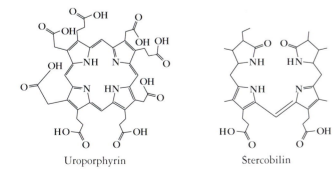

Uroporphyrin   Stercobilin

Whereas we excrete our urine, the Turacos (this one is *Musophaga rossae Gould*) wears it on its head. The red head feathers are colored by a molecule closely related to the one that colors our urine.

PUTRESCINE (131) C$_4$H$_{12}$N$_2$

The names of this amine and its companion *cadaverine,* which differs by the addition of another —CH$_2$—group to the chain, speak for themselves; little more need be said to describe their odor or to indicate their origin in rotting flesh. However, they also contribute to the odors of living animals and are partly responsible for the smell of semen. Both add to the odor of urine and are present in bad breath (exhaled air that, sometimes in the lungs, has picked up volatile compounds from the blood). Putrescine is a poisonous solid and cadaverine a poisonous, syrupy liquid; both have disgusting odors.

We dress in the odor of death. One of the monomers in nylon-6,6 is hexamethylenediamine (71), a diamine with six —CH$_2$— groups. However, we cannot smell death in a nylon garment, partly because the amine molecules are anchored to their neighbors and cannot reach the nose.

TRIMETHYLAMINE (132) C$_3$H$_9$N

The trimethylamine molecule is a derivative of the ammonia molecule (7) in which there are three —CH$_3$ groups in place of ammonia's three hydrogen atoms.

Trimethylamine is a gas (but condenses to a liquid at 2°C). It has the odor of rotten fish. Indeed, when you smell rotting fish, this is the molecule you can think of as being inside your nose, for it is given off as it is formed by enzyme and microbial attack on fish proteins. The enzymes released during gutting are particularly potent at carving proteins into trimethylamine. This amine is also secreted by the coyote and the domestic dog, which also sometimes smell of rotten fish.

The saffron crocus (*Crocus sativus*).

5

S I G H T

A N D   C O L O R

Communication between an object and an eye is physical, since the message is carried by a light ray. However, molecules are often the agents of color in an object. They are also the sensors that respond to light once it has entered our eyes.

Many natural colors depend on the presence of particular molecules, and the following pages show you some that inhabit petals, leaves, and skin and are responsible for the colors of flowers, trees, and people. Now, when you look upon a rose, you will know the molecules that contribute to its shades and hues. You will discover what changes give rise to the colors of a leaf in the fall, and you will comprehend the ebb and flow of its color molecules. You will also understand the events that take place in your eye when you perceive these colors. You will see that, in vision, molecule responds to molecule, and that the activated molecule in the eye triggers a signal that, in the deep unknown of the head, is detected as color. You will also learn what molecules contribute to the colors of hair, meat, and fat, and how some people do away with their coloration while others seek to enhance theirs naturally or artificially.

We begin with the perception of color and then move on to the molecules responsible for various colors. Throughout the chapter, you should keep in mind the idea that many substances respond to light because their electrons can be shifted by the energy that light brings.

VISION

Although the receptors in the eye are sensitive to a physical stimulation—light of various wavelengths—once light has struck, all the detection, processing, and transmission of the neural signal are chemical. The retina of the eye consists of two kinds of receptors, called *rods* and *cones*. The billion or so rods function in dim light but do not distinguish among colors (light of different wavelengths). The 3 million or so cones function in bright light and do distinguish between colors. In fact, there are three kinds of cones, each of which absorbs either red or green or blue light and sends signals to the brain accordingly. Each receptor contains molecules that are sensitive to light, and their response to illumination is the trigger for a message to the brain.

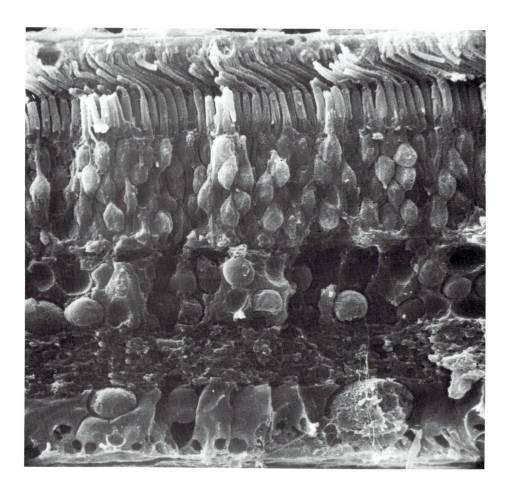

A scanning electron micrograph of the retina of a rabbit. The rods and cones are at the top and the ganglia are the spherical bodies at the bottom.

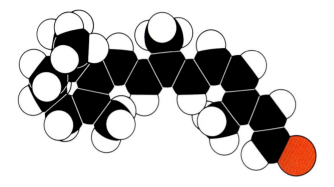

CHO

The retinal molecule has a shape that will recur as a structural theme in this and the next section, for reasons that will soon become clear. The molecule is largely hydrocarbon, and its most important feature is that it consists of an alternating chain of carbon-carbon single and double bonds. That chain terminates in a —CHO group typical of *aldehydes (hence the -al in its name).

The alternating double and single bonds result in two properties. One is the rigidity of the tail: The hydrocarbon chain is not free to wind up into a coil like a singly bonded hydrocarbon (20), but sticks out rigidly from the six-member ring. The other consequence is that the electrons in the chain are only loosely held and can be moved to new locations reasonably easily. Hence, the molecule can absorb energy from light that falls on it, storing that energy as a shift of its electrons into a new arrangement. The —CHO group at the end of the chain is important insofar as it is quite reactive and can combine with (and anchor the retinal molecule to) other molecules in its vicinity—particularly protein molecules.

The 11-*cis*-retinal molecule is the one that absorbs incident light in the rods and cones of the eye. In the rods, the retinal molecule is linked to a protein, *opsin,* to give the light-sensitive *rhodopsin.* Rhodopsin is also called *visual purple.* In the cones, retinal molecules are linked to three slightly different opsins that change the wavelength of the light they can absorb to red, blue, and green. Retinal can also be excited by ultraviolet light, but this is normally filtered out by a yellow pigment in the cornea, and we are confined to seeing by longer-wavelength "visible" light. Some people who have had cataracts removed, however, can read by ultraviolet light.

It should not be surprising that the light-absorbing ability of retinal and its protein combination rhodopsin are used to extract energy from sunlight, and not merely for vision. The purple, nonsulfur bacterium *Halobacterium halobium,* which is commonly found in sunny, concentrated brine that is about seven times more salty than sea water (as at the edges of salt pools), uses it (in the form of bacteriorhodopsin) as space stations use a solar cell. The halobacterium uses the energy the retinal captures to empty its cell of hydrogen ions, which leads to the formation of ATP (98) from ADP, which in turn powers the activities that constitute its little life.

trans-RETINAL   (134) $C_{20}H_{28}O$

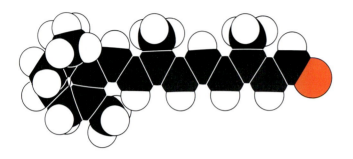

The *trans*-retinal molecule is identical to the *cis*-retinal molecule (133) with the exception that one half of the molecule has been twisted into a new rigid arrangement around one of the double bonds, so that the side chain is fully extended (and still rigid). (See page 10 for a discussion of cis and trans.)

When light strikes a rod or cone, it is absorbed by the *cis*-retinal in that receptor. As in all molecules this light absorption causes a shift of an electron. In this case, one of the two electron pairs between the two atoms where the *cis*-retinal molecule is bent at a right angle is split apart; that is, the double bond is suddenly replaced by a single bond. One end of the molecule is now free to rotate relative to the other, and *cis*-retinal suddenly changes its shape to *trans*-retinal. Once it has done so, the two electrons of the divided pair come together again, re-form the double bond, and freeze the molecule into its new shape. This massive change of shape affects the shape of the opsin protein and causes a signal to be sent along the optic nerve to the brain. After that has

happened, the *trans*-retinal breaks away from its location on the opsin, where it no longer fits, is converted elsewhere back to *cis*-retinal, and then reattaches to an opsin to await its next illumination.

All image-resolving eyes (which are found only in the mollusks, arthropods, and vertebrates) have adopted retinal as their light-sensitive component, even though they may have undergone separate evolution. The halobacteria mentioned above, although of a different kingdom, also have a vestigial light-sensing ability based on retinal. All these creatures have adopted the molecule because it is ideal for its purpose. It undergoes a large change of shape when it isomerises from cis to trans, and its absorption wavelength is readily and widely modified by its protein neighbor. Moreover, the cis form is structurally stable and does not readily convert to the trans in the dark (so that we are not misled and confused by false signals). It can also be synthesized from readily available precursors, including the carotenes (136).

LEAVES, CARROTS, AND FLAMINGOS

The colors of many substances, including those described in this and the next section, are due to *absorption* and must be distinguished from colors due to emission (page 17) and *incandescence. You need to know two facts to understand how absorption colors arise.

One is that white light is a mixture of all possible colors: On the sun, the incandescence of the surface layers results in the mixture of *wavelengths that we are accustomed to as "white" light. If any color is removed from white light, then the light takes on a hue. Filtering orange light out of white light, for instance, results in blue-green (cyan) light; filtering out cyan results in orange light. The color resulting from the removal of a color from white light is the latter's *complementary* color. A traditional way of demonstrating the relation between colors is in terms of an artist's *color wheel,* on which complementary colors are opposite each other along a diameter.

The second important point is that substances can absorb only certain wavelengths of light. If their electrons can be rearranged to states that differ only a little in energy from the normal arrangement, then they can absorb red light, since red (long-wavelength) light is low-

energy light. If the incident light is white, then the light reflected by such substances is perceived as green. If a substance needs to absorb a lot of energy for its electrons to be changed into a new arrangement, then it absorbs only short-wavelength light, since short wavelengths correspond to high energy. If it absorbs blue-indigo light (which is short-wavelength light), then the light it reflects is perceived as orange.

The color green is interesting in this regard. On the color wheel, it lies opposite both red and violet, where the two ends of the visible spectrum meet. Hence it can arise when low-energy, long-wavelength red light is removed from white light, when high-energy, short-wavelength violet light is removed, or when both are removed.

The bright reds, yellows, and blues of flowers are due to a class of compounds treated in the next section. But many of the yellow and orange hues of nature are due to molecules called *carotenoids,* which we will consider here. When you examine their structures, it would be good to keep in mind their common structural unit, for carotenoids (and parts of chlorophyll too) can all be regarded as mainly strings of isoprene units (62). It should hardly be

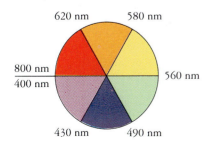

Complementary colors are diametrically opposite each other in an artist's color wheel. Thus, if a substance absorbs red light from incident white light, the reflected light is the complementary color of red, which is green.

surprising that rubber is exuded by trees, since rubber is a polyisoprene (63), or that humans emit it (page 140), for vitamin A is a chain of isoprene units. Many plant odors [particularly the terpenes (page 132)] are also built from isoprene units—emphasizing again the economical elegance of nature.

Here, though, we are concerned with color and carotenoids: The color of every carotenoid molecule is due to the ability of light to break open loosely held pairs of electrons in the extended chain of alternating single and double bonds. One of these electrons is raised to a higher energy state and, in the process, light of a particular color is absorbed.

Carotenoids can be passed from a species to its predator, and that, as you will see, leads to some interesting relations.

CHLOROPHYLL   (135) $C_{55}H_{72}MgN_4O_5$

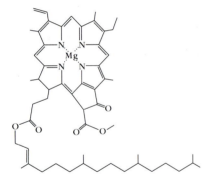

The ubiquitous green of leaves is due to *chlorophyll,* the structure of which is shown in the illustration above. The atom at the center is magnesium.

This hugely important molecule absorbs both violet and red light, which is exactly what is required for the reflected light to appear green. Hence vegetation is green. The absorbed light is the energy supply for the task of *photosynthesis,* in which carbon dioxide (4) and water (6) are combined to form carbohydrates (79). The molecule of chlorophyll is thus the antenna that green plants use to harvest the sun and open the way to the processes of life.

Chlorophyll absorbs very strongly and can mask many other colors. Some of these are unmasked when the chlorophyll molecule decays in the fall (page 157). When vegetation is cooked, the central magnesium atom is replaced with hydrogen ions. This changes the energy needed to excite the electrons in the rest of the molecule, so that cooked leaves change color—sometimes becoming an insipid green.

CAROTENE (136) C$_{40}$H$_{56}$

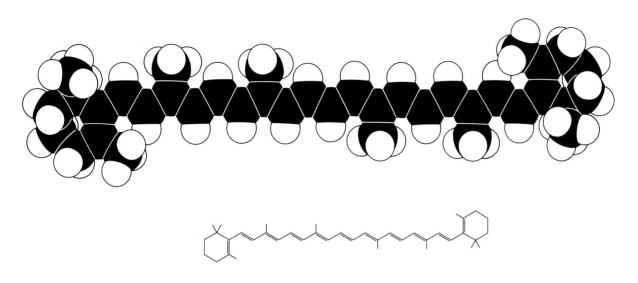

The carotene molecule is a hydrocarbon and a string of eight isoprene (62) units. We take it as the starting point for all the other molecules in this section.

The most striking feature of the carotene molecule is its chain of alternating carbon-carbon single and double bonds. As explained in the discussion of the closely related molecule retinal (133), this arrangement has two consequences. One is that the molecule is stiff and inflexible. The other is that the electrons in the extended chain are only loosely held and can readily be excited by low-energy light: Carotene itself can absorb blue-indigo light and hence looks orange. Another feature of the molecule relevant to its occurrence in the world is that, being a hydrocarbon, it is soluble in fats, which provide a similar oily environment, but not in water.

Carotene occurs in carrots, but their yellow core is due to slightly oxidized versions of carotene known as *xanthophylls*. In fact, our familiar orange carrots are a fairly recent innovation, for their precursors were purple, owing to the presence of molecules like those described in the next section. However, a mutation led to a variety in which that pigment did not develop, leaving the golden yellow carotene molecule dominant. Carotene is also partly responsible for the color of the mango and the persimmon. The pale cream of milk and the yellow of butter are due largely to the presence of molecules related to carotene and which are added to margarine to emulate the color of butter. The fat of meats is often tinged slightly yellow by carotene that the animal ingested and which, by virtue of its hydrocarbon nature, dissolved in the fat.

Carotene accompanies chlorophyll in photosynthetic organisms. Its role is partly to harvest some sunlight that is not absorbed by chlorophyll, as well as to react with energetic oxygen molecules so as to protect the cells from degradation. There is typically about one carotenelike molecule to every three chlorophyll molecules in a leaf, so that the darker the green of the leaf the greater the concentration of carotenes. The yellow-orange of carotene remains masked by the chlorophyll until the fall, when the chlorophyll molecule decays and is not replaced; that leaves the sturdier carotene molecule [with other color molecules (page 157)] to exhibit its powers of light absorption, and the leaves turn yellow.

The persimmon owes its color to carotene and its astrigency to tannins. The latter become insoluble as the fruit ripens and the bitterness lessens. The persimmon tree is a member of the ebony family (*Ebenaceae*).

When grass is cut and bleached in the sun, its carotene molecules are broken down into molecules of ionone (109), which has the characteristic odor of hay.

Lycopene is, in essence, a carotene molecule with its two rings broken open. It is a deeper red than carotene and is responsible for the color of tomatoes *(Lycopersicon esculentum).* As a green tomato ripens, the chlorophyll of the unripe fruit decays, the increasing amount of lycopene is unmasked, and the fruit turns red. Lycopene and carotene join forces to contribute color to apricots.

A shorter version of the lycopene chain, with each end partially *oxidized to become a *carboxyl group, is the molecule *crocetin.* This molecule is responsible for the color of *saffron* (from the Arabic word *za'faran,* meaning "yellow") and is used, among other applications, to color saffron rice. Saffron is obtained from the stigmas of the eastern crocus *(Crocus sativus).*

A close relative of crocetin, differing from it by way of a slight lengthening of the chain and the conversion of one carboxyl group to its methyl ester, is *bixin.* Bixin is the molecule responsible for the red color of *annatto,* which is obtained from the pulpy parts of the seed of the tropical shrub *Bixa orellana* and used to color cheeses such as red Leicester.

The colors of the petals of the Indian Blanket (*Gaillardia pulchella*), which grows widely in Texas, result from a yellow carotenoid molecule and a magenta anthocyanin that is normally absent from the tip. These combine to give the brick-red hue of the rays.

ZEAXANTHIN (137) $C_{40}H_{56}O_2$

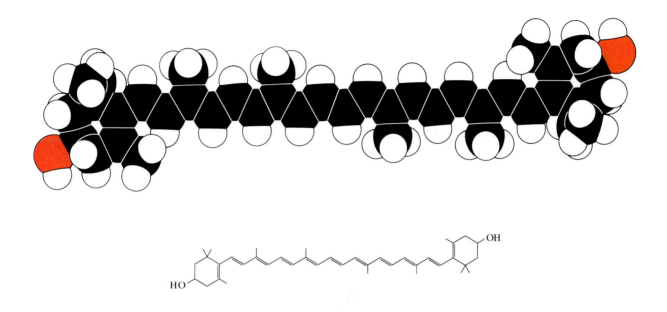

With zeaxanthin we move to the oxygen-containing *xanthophylls* and to the corn belt, for it and its relatives occur widely and make great stretches of the world golden. Zeaxanthin and carotene jointly color corn *(Zea mays)* and contribute to the color of the mango.

Zeaxanthin is also a partner in the coloration of egg yolk and orange juice, where it is joined by *lutein,* a molecule that differs from zeaxanthin only in the location of the last double bond on the right. When they are ingested, lutein and zeaxanthin dissolve in hydrocarbons (because they are largely hydrocarbon themselves) and contribute to the yellowish tint of animal fats.

The yellow corn takes its color from zeaxanthin; this cob shows kernels obtained from self-pollination (the yellow) and cross-pollination from a purple stock. The color of the latter is due to an anthocyanin that masks the zeaxanthin.

ASTAXANTHIN   (138) $C_{40}H_{52}O_4$

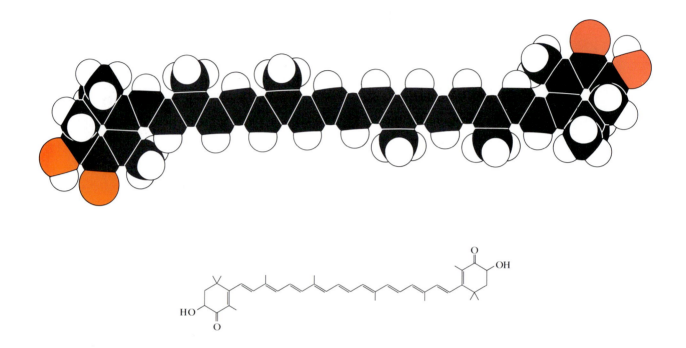

The presence of four oxygen atoms in astaxanthin changes the energy needed to move its electrons, so that its color is different from that of carotene. Astaxanthin is, in fact, pink (despite its name, which comes from the Greek words for "yellow flower") and is responsible for the color of salmon. Astaxanthin also occurs in the carapaces of shellfish, including lobster and shrimp; however, its pink color is not apparent in the live animals because the molecule is wrapped in a protein, giving a blackish hue. When lobster and shrimp are boiled, the protein chain uncoils, liberating the astaxanthin molecule, and the lobster and shrimp turn pink.

An astaxanthin molecule that has lost its two —OH groups is called *canthaxanthin*. That molecule is responsible for the color of the American flamingo, which owes its color to its diet: Flamingos in captivity lose their pink color if they are not supplied with adequate amounts of carotenoid-containing shrimp.

Shrimp after they have been cooked. The color change is a result of the release of astaxanthin from a combination with a protein.

FLOWERS, FRUIT, AND WINE

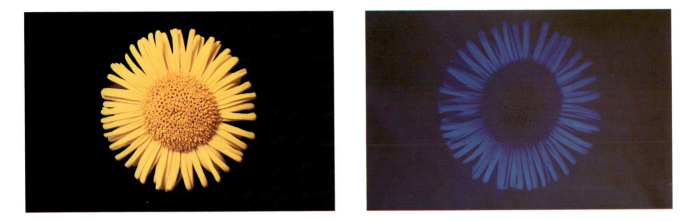

To our eyes the fleabane flower (*Pulicaria dysenterica*) is yellow, but to eyes sensitive to ultraviolet light its color would be "bee's purple."

Many of the bright colors of the world in spring, summer, and fall are due to a single class of compounds called *flavonoids*. They have in common a basic framework like that shown in the line formula below, which

consists of two benzenelike rings and another that includes an oxygen atom. Most of them occur in combination with a sugar molecule, producing a *glycoside, but the sugar component will not be shown in the models. Flavonoids occur in leaves as well as in petals. Their function in leaves is to absorb the ultraviolet light that could be so destructive to the genetic material and the proteins in the cell.

Nature mixes its palette of colors from the framework of flavonoids by attaching different groups to different places around it, linking different types of sugar molecules to it, and changing the acidity of its environment.

Eyes other than our own can see a richer palette. Bees, for instance, have eyes that are sensitive to *ultraviolet light, so they see a richer range of colors where we might see only one. An example is the (to us) plain yellow of the fleabane (*Pulicaria dysenterica*). To the bee, which notices both the ultraviolet and the yellow that are reflected, these flowers appear a color called "bee's purple."

Ultraviolet light absorption can occur accidentally in animals that have digested flower pigments and can induce distress. When cattle feed in pastures that includes the yellow St. John's wort (genus *Hypericum*), the coloring pigment finds its way to the surface of the skin, where ultraviolet light is absorbed and initiates a chemical reaction that inflames the skin and, in severe cases, can lead to death.

PELARGONIDIN (139) C₁₅H₁₁O₅

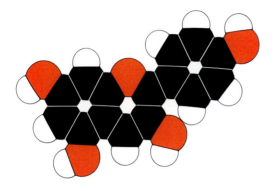

Pelargonidin is the simplest of the *anthocyanidins,* the principal class of flavonoids discussed here. In combination with sugar molecules such as glucose (79), anthocyanidins become *anthocyanins* (from the Greek words for "blue flower"). Anthocyanins are responsible for many of the red, purple, and blue colors of nature (but not the red of either beet or *bougainvillea,* which is due to other types of compounds called *betacyanins*). Pelargonidin itself is responsible for the red of the common geranium (genus *Pelargonium*) and contributes to the color of ripe raspberries and strawberries.

Cyanidin, an anthocyanidin with two —OH groups on the right-hand benzene ring, provides the violet on the palette; it is responsible for the color of ripe blackberries and contributes to the colors of black currants, raspberries, strawberries, apple skins, and cherries. Its color is responsive to the acidity of its surroundings: In acid solution it is red, but in *alkali blue. Here nature again shows its economy, for the strikingly different colors of the blue cornflower (*Centaurea cyanus*) and the red poppy (*Papaver rhoeas*) actually arise from the same molecule (see page 3). In the cornflower the sap is alkaline, and the cyanidin molecule loses a hydrogen ion and turns blue. In the poppy the sap is acid; because the molecule is in an environment rich in hydrogen ions, it acquires

one and turns red. Rhubarb, which is rich in oxalic acid (90), is colored red by the acid form of cyanidin. Red cabbage retains its color, which is due to cyanidin, if it is kept acid as it is cooked. Flowers sometimes modify the acidity of their sap and change color after pollination so that they become less conspicuous to insects.

Replacing an —OH group of the cyanidin molecule with an —O—CH₃ group brings us to the *peonidin* molecule and the color of the peony (genus *Camelia*). Peonidin also contributes to the color of cherries and grapes.

The colors of spring, summer, and fall are found in a glass of wine. Red wine acquires its color from anthocyanidins. As it matures, the anthocyanidins react with other largely colorless but bitter flavonoids that are also present and are known collectively as a type of *tannin.* This reaction removes the tannins and improves the taste of the wine. As red wine grows even older, the reaction between the anthocyanidins and the tannins results in the removal of the red anthocyanidins and leaves the brown tannins visible. The color of white wines is due in part to quercetin (140), which turns deeper brown as more is *oxidized with age. Initially, a young white wine may have a greenish hue from the chlorophyll (135) molecules that survive fermentation.

QUERCETIN (140) C₁₅H₁₀O₇

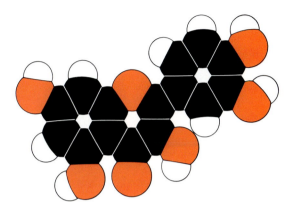

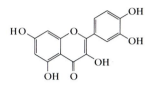

The quercetin molecule is a flavonoid, but it is a *flavonol* rather than an anthocyanidin (the extra oxygen on the ring makes the difference). This molecule is yellow and is responsible for the color of Dyer's oak *(Quercus tinctoria)*. Quercetin also occurs in many leaves, but its color is masked by chlorophyll (135) until the latter decays in the fall.

Some flavonols are colorless (to our eyes) and are present in leaves to prevent damage by absorbing ultraviolet radiation. They do not absorb blue and red wavelengths and hence do not intrude on the energy-harvesting function of chlorophyll. In the fall, when the chlorophyll decays, these colorless flavonols are stripped of the oxygen atom attached to the ring and thus are converted to anthocyanidins such as the scarlet pelargonidin (139). This switch of compound, induced by no more than the loss of a single oxygen atom, is what we perceive as the majestic blaze of autumn glory.

If the quercetin molecule is modified by removing the —OH group on the heterocyclic ring, the *luteolin* molecule is obtained. Luteolin is an example of a *flavone,* another common type of pigment molecule. It is the yellow dye of the chrysanthemum.

The colors of the carotenes, anthocyanins, and flavonols in leaves are masked by chlorophyll, but when that decays in the fall their yellows, oranges, and reds are revealed.

BROWNS, BRUISES, AND TANS

Not everything is brightly colored, and some things that are fresh and white become brown when bruised. Many people are brown even before they are bruised. Hair, unless it courts fashion deliberately and synthetically, is rarely brightly colored.

Some browns are formed by reactions between carbohydrates (79) and the amino acids (73) of proteins. Such a combination, which is called a *Maillard reaction,* generally occurs only when the two substances are heated, and it results in a very complex mixture of products. At the same time as the brown color develops, molecules that cause taste and odor are formed, so that browning induced by heating contributes to flavor. Maillard reactions occur on the hot, dry crust of bread while it is being baked, and some of the more mobile flavor molecules diffuse into the interior of the loaf. Browning reactions also occur when maple syrup is boiled, as the sugar and amino acids in the sap combine: The longer it is boiled, the more extensive the reaction and the deeper the color of the syrup. The color of beer also darkens during brewing, as the unfermented sugar reacts with the amino acids present. Roasting coffee and cocoa beans and nuts of various kinds also brings out the flavor as well as the color by stimulating Maillard reactions between their carbohydrate and protein components.

Another familiar brown is that of *caramel,* formed when pure sucrose (81) is heated. This is not a Maillard reaction, for no amino acids are present. Instead, it is the result of very complex decomposition and recombination reactions caused by the heat. Small, tasty, smelly molecules, including acrolein (102), are also produced by the decomposition and contribute to the flavor of caramel.

Many of the browns and blacks of nature can be traced to a single molecule—*melanin*—the one we consider first.

· ·

TYROSINE (141) $C_9H_{11}O_3N$

MELANIN (142)

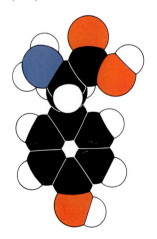

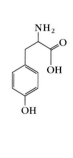

(141)

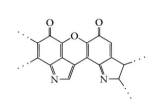

(142)

Tyrosine is a naturally occurring *amino acid that is part of many proteins. It is also the starting point for the synthesis, in the body, of some of the molecules responsible for transmitting signals from one nerve cell to another. For our purposes, though, it is the start of the trail that leads to melanin.

Under the action of enzymes, tyrosine molecules *polymerize to give the melanin molecule, a part of which is shown in the line formula on the opposite page. The electrons in the alternating double and single bonds are only loosely held and can be moved around by light of any wavelength. Hence, all light is absorbed by melanin and an object that contains it looks black. Melanin molecules become attached to protein molecules and accumulate in granules that range in color from yellow through brown to black.

Melanin contributes to the pigmentation of skin and hair [except that of redheads, where the iron-rich pigment trichosiderin is dominant, and of those who have destroyed the double bonds by *oxidation with hydrogen peroxide (14)]. It also provides a dark background that makes colors due to light scattering more obvious, as in the iris of the eye. The number of melanin-producing cells is much the same in dark- and light-skinned people, but they are more active in dark-skinned people.

Melanin is a part of the color-change mechanism of the chameleon, in which it is shipped around through channels in the skin and used to cover the brighter pigments below. Animals that darken themselves, including the octopus and its relatives, achieve the change by dispersing granules of melanin; the skin lightens when the melanin granules aggregate again.

Melanins with slightly different compositions are also formed when fruit is bruised; their formation follows damage to cell walls that allows an enzyme, *phenol oxidase,* to act on the material within. This enzyme is absent from citrus fruits, melons, and tomatoes, which do not brown so rapidly when bruised. However, the precursors of melanin are not quite the same in fruit as in animals, for although they contain, like tyrosine, an —OH group attached to a benzene ring (that is, they are *phenols), they are not necessarily amino acids.

The dark colors of tea are due to similar melaninlike polymerized phenols.

The octopus (*Octopus vulgaris*) is one animal that deploys melanin to change its color.

AMYL *para*-DIMETHYLAMINOBENZOATE   (143) $C_{14}H_{21}O_2N$

DIHYDROXYACETONE   (144) $C_3H_6O_3$

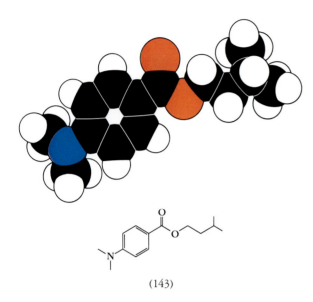

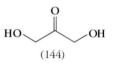

(143)

(144)

In addition to reddening of the skin, the ultraviolet light in sunshine causes two types of tanning. One form occurs immediately (and continues in corpses) but disappears within a few hours; it is due to damage to the molecules from which melanin is formed. A more long-lasting tan (which depends on biochemical processes and does not occur in the dead) is induced by ultraviolet light in a narrower range of wavelengths. This tan is due to increased activity of the melanin-producing cells deeper in the skin; their product becomes apparent when the melanin has had time to reach the surface, normally within a day or so. The melanin acts like the flavonoids of plants (139), protecting the DNA in cells from damage.

The aminobenzoate is typical of the molecules contained in *sunscreens;* they are included to absorb ultravio-let light of wavelengths that the ozone layer (5) is ineffectual at blocking. A suntan cannot be achieved without sunburn, because sunburn triggers melanin production. Sunscreens act by lowering the dose of damaging radiation received by the skin, so that melanin production has a chance to catch up with sunburn.

Quick-tanning lotions cut out all the radiation that causes sunburn (and hence true tanning), permitting only the immediate-response, broad-spectrum, destructive tanning. Their effect is often enhanced by the inclusion of a skin dye such as dihydroxyacetone, which reacts with the amino acids in the upper layers of the skin and colors them brown through a version of the Maillard reaction used for browning foods. This tan is ephemeral, for it is lost as the stained cells are rubbed off.

Ripe coffee beans and flowers.

6

THE LIGHT AND THE DARK

One of chemistry's major contributions to the welfare of humanity has been its provision of pharmaceuticals. We know how some of these drugs act, and it is possible to discern a relation between the structure of their molecules and the influence they have on our bodies. Some of them quell pain. Others induce calm or expunge a depression. Still other molecules do the opposite and induce a sense of euphoria, sometimes simultaneously inducing addiction and, through addiction, ruination. Unhappily, chemists achieve evil as well as good, sometimes by accident but sometimes by intention. It would be improper to conceal this dark face of their activity, to show their benevolent creations but not their malevolent, so a few of their pernicious contrivances are described here.

On the brighter side, chemists are helping biologists to decipher the secrets of life by unraveling the strands of simplicity that shroud the complexity of life. Chemists now have the skill and technology to trace deep things. As an epitome for this impressive elucidation of nature, I include a couple of molecules to titillate the imagination. Through them you will see that men and women differ, at their molecular roots, by little more than a methyl group!

PAIN KILLERS AND TRANQUILIZERS

Analgesics can be classed in two broad categories. One consists of those that act peripherally, at the site of the pain, to interfere with the pain signals at their source. The other group acts centrally, on the central nervous system, to modify the brain's processing of the signals it receives along its pain nerves (page 114). Centrally acting analgesics often produce disadvantageous side effects, such as addiction and mood modification. Peripherally acting analgesics are not addictive; nor do they modify mood directly.

A number of the milder nonaddictive analgesics on the market are aspirinlike. They are often derivatives of salicylic acid, whose name comes from the Latin name of the white willow, *Salix alba:* The acid appears in the bark of that tree, in combination with a sugar molecule (79), as the *glycoside *salicin.* Extracts of the bark were originally used to relieve pain, as in oil of wintergreen (124). Extracts of the leaves and bark of a number of trees and shrubs have similar medicinal properties that can be traced to the same substance or close relatives.

. . . .

SALICYLIC ACID   (145) $C_7H_6O_3$

ACETYLSALICYLIC ACID   (146) $C_9H_8O_4$

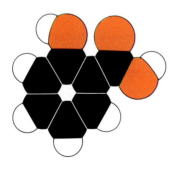

(145)

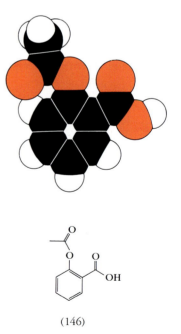

(146)

Salicylic acid is the parent compound of a family of analgesics, particularly acetylsalicylic acid, which is commonly known as *aspirin* (a generic name in most countries, but a trade name in some). Aspirin was discovered in 1899, but its mode of action remained unclear for a long time. It is now thought to interfere with the synthesis of *prostaglandins* by inhibiting the action of an enzyme, *prostaglandin cyclooxygenase*. Prostaglandins are locally acting hormones that are involved in numerous processes in the body; among them is the modification of signals transmitted across synapses (the connections between nerves), particularly pain signals.

Prostaglandins may also be involved in the dilation of blood vessels that causes the pain experienced as a headache (if the vessels are intracranial) or as migraine (if they are external to the skull). In these cases, aspirin and other local analgesics may act by inhibiting the synthesis of the prostaglandins that might cause the pain as well as help transmit it. The zigzag lines often seen by victims of migraine are caused by constriction of the blood vessels in the region of the brain responsible for vision. Since they have a different cause from the headache, they can be treated independently with small doses of substances (including amyl nitrite) that can dilate the vessels.

The proprietary formulation *Alka-Seltzer* consists of aspirin, anhydrous citric acid (91), and sodium bicarbonate ($NaHCO_3$). When the mixture is added to water, the aspirin (an acid) is converted to its more soluble sodium salt, and the citric acid releases carbon dioxide (4) from the bicarbonate ions, giving the fizz as well as a pleasant taste.

....

N-ACETYL-*para*-AMINOPHENOL   (147) $C_8H_9O_2N$

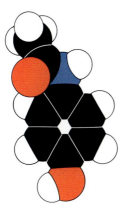

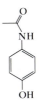

The analgesic properties of drugs such as that shown were discovered by accident when some *acetanilide* (like the molecule in the illustration, but without the —OH group) was added by mistake to a patient's prescription. Acetanilide can, however, be toxic, and less harmful compounds were sought. One of them is *N*-acetyl-*para*-aminophenol, better known as *paracetamol* and sold under the name *Tylenol*. Acetanilide is, in fact, converted into paracetamol in the body, which gives it its analgesic properties; but some is also converted into *aniline,* a benzene molecule with one hydrogen atom replaced by an —NH₂ group, which gives it its toxicity.

The marked similarity between this molecule and acetylsalicylic acid (146)—aspirin—should be noted. Although they are built up from different atoms they have similar shapes, with an —NH— group in the aminophenol taking the place of an —O— in aspirin. Because of their similarity they are both recognized by the same enzyme, the one responsible for the biosynthesis of prostaglandins, and paracetamol also acts by inhibiting prostaglandin production.

....

DIAZEPAM (148) $C_{16}H_{13}ON_2Cl$

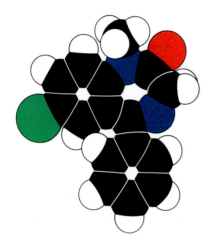

Much insomnia is due to anxiety, which is associated with heightened neuronal activity in the *limbic system, the region of the brain associated with emotion. This activity can spread to the *brain stem*, the most primitive part of the brain, and maintain wakefulness.

The molecules called *benzodiazepines* bind to protein molecules in the junctions between nerves and enhance the ability of the neurotransmitter GABA (28) to bind to neighboring sites on the same molecule. As explained in the discussion of the depressant action of alcohol (page 44), GABA inhibits the firing of nerve cells, so that the presence of the benzodiazepine molecule encourages this inhibition. The sites to which benzodiazepines bind are particularly rich within the limbic system, so ingesting a benzodiazepine suppresses the abnormal activity there that we experience as uneasiness and generalized fear and know as anxiety. That is, benzodiazepines are specifically antianxiety drugs that (in larger doses) act indirectly as sedatives.

Benzodiazepines are slightly addictive but not particularly lethal. However, ethanol and the benzodiazepines bind at neighboring sites on the same protein, and the two together can so change the shape of the protein that there is a strongly enhanced propensity for GABA to bind. This results in a massive inhibition of nervous activity and may lead to death.

The benzodiazepines include the diazepam molecule illustrated here which is widely sold under the trade name *Valium* and used as a muscle relaxant as well as a tranquilizer. If the chlorine atom on the left of the diazepam molecule is replaced by an —NO$_2$ group, and the —N—CH$_3$ by —NH, the result is *nitrazepam*, which is sold as *Mogadon*. Another tranquilizer, *chlordiazepoxide*, has a very similar structure and is sold as *Librium*.

MORPHINE   (149)   C$_{17}$H$_{19}$O$_3$N

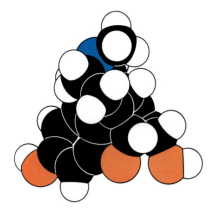

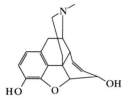

Morphine is the principal component of *opium* (from the Greek word *opion,* for "poppy juice"), which is obtained as the milky juice that exudes from unripe poppy seed capsules *(Papaver somniferum).* (See the photograph on page 166.)

Unlike the aspirin analgesics, morphine and related compounds act on the central nervous system and can induce addiction. The specific action of morphine and its relatives appears to be related to the ability of the molecule to fit into and block a specific receptor site on a nerve cell. The shape of the receptor that has been pro-

posed is shown in the illustration below. The benzene group of the morphine molecule fits snugly against the flat part, and the neighboring group of carbon atoms is at the correct distance and orientation to fit into the groove. Beyond the groove is a group with a negative charge, which can attract the positive charge of the nitrogen atom. By fitting the shape of the receptor and binding to it, the incoming morphine molecule eliminates its action. In this respect the molecule mimics the body's natural pain killers, the enkephalins (page 114).

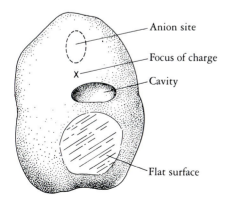

The supposed shape of the protein that acts as the receptor site for morphine and related molecules.

The opium poppy *(Papaver somniferum)* from which opium is obtained. The heads are seen on the right.

Morphine acts on the deep, aching pain described on page 114 as *slow pain,* but has no effect on fast pain. Large accumulations of morphine receptor sites are found· in the substantia gelatinosa, the region of the spinal column where pain signals are first processed (page 114); there, morphine acts by raising the threshold at which slow pain is experienced. Morphine receptor sites are also abundant in the medial region of the thalamus, the part of the brain that acts as an input region for slow-pain signals. High concentrations are also found in the limbic system, which, as we have noted, is a region closely associated with emotion. This point will become relevant when we consider the stimulation of euphoria in the next section.

A close relative of morphine is *codeine,* which is obtained when the —OH group on the left in the illustration is replaced by —O—CH₃. In the body, the —O—CH₃ group is replaced by —OH, so that codeine is converted back to morphine.

A *heroin* molecule is a morphine molecule in which the hydrogen atoms of two —OH groups have been replaced by acetyl groups (—CO—CH₃). Other names for heroin are *diacetylmorphine* and *diamorphine.* The replacement of the hydrogen-bonding —OH groups results in heroin being less soluble in water than morphine but more soluble among the hydrocarbon chains of fats (36). Therefore, although it must be injected directly into the blood, it passes more rapidly through the blood-brain barrier, the barrier that prevents water-soluble and large molecules from passing between the two. As a result, it is more potent than morphine (more "heroic"), but its effect does not last as long. Once heroin is absorbed into the body, the acetyl groups are removed, forming morphine, which provides its analgesic and euphoric action.

STIMULANTS

Many drugs are used for nonmedical purposes in ways that vary in their degree of social acceptability. This section describes a few that are used in one way or another to bring about a sense of euphoria. Ethanol, a depressant (27), has already been discussed.

As we have noted in several previous sections, the region of the brain associated with emotion is the *limbic system: Stimulation of the neurons of the limbic system, either by enhancing the ability of neurotransmitters to communicate between them at synapses or by depressing the ability of the inhibitory transmitter GABA (28) to bind, can result in enhanced emotional activity. Now we need to bring the story slightly more into focus.

Many of the connections in the limbic system involve highly branched neurons that originate in a tiny blue region of the brain stem called the *locus coeruleus*. The chemically significant feature of these neurons is that

their neurotransmitter is a molecule called *norepinephrine* (or *noradrenaline*); this neurotransmitter is also abundant in the limbic system itself. The neurophysiologically significant feature of these neurons is their branching, which results in numerous synaptic connections with other neurons, both in the limbic system and in the higher regions of the brain. It seems reasonable to assume (but nonetheless still speculative) that such highly branched, polysynaptic neurons correlate with emotional responses, whereas less branched, even monosynaptic, neurons correlate with more logical, rational functions. If so, then any modification of the response of these highly branched clusters of neurons can be expected to modify emotional responses. The key to that modification may be their neurotransmitter, the molecule norepinephrine.

. . . .

ADRENALINE (150) $C_9H_{13}O_3N$
AMPHETAMINE (151) $C_9H_{13}N$

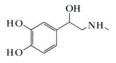

(150)

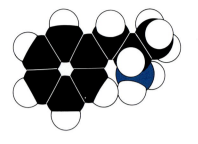

(151)

Not all stimulants need to be administered from outside; some, including adrenaline, are produced within the body and participate in the transmission of signals from one nerve cell to another. Adrenaline is produced in the renal gland of the kidney. It is also called *epinephrine*, a Greek version of its name (*ren* is the Latin word for "kidney," *nephros* the Greek). Adrenaline circulates in the blood and affects the autonomic nervous system (the nerves over which we do not have voluntary control, as distinct from the central nervous system). It acts on the heart muscles to increase their force of contraction, it dilates the pupils of the eye, and it stimulates the secretion of sweat and saliva. However, the *hydrogen-bonding —OH and —NH— groups in the molecule keep it from passing across the hydrocarbon-rich curtain of the blood-brain barrier.

Nevertheless, adrenaline and closely related compounds can be made in the brain. One of these related compounds is the norepinephrine mentioned in the introduction to this section. Its precursor is the amino acid tyrosine (141), which has the same basic ground plan as adrenaline and is converted into norepinephrine by enzymes. This conversion occurs abundantly in the neurons of the locus coeruleus and limbic system and, in a sense, feeds our emotions.

Now consider the amphetamine molecule. Amphetamine, which was once widely available under the trade name *Benzedrine*, is a stimulant. The similarity of its shape to that of norepinephrine suggests that it can act in a similar manner, to stimulate the limbic system and the locus coeruleus. Since the latter is also connected to the higher centers (particularly the cerebral cortex), amphetamine may also stimulate the higher, cognitive functions and lead to greater alertness and a feeling of prowess.

The action of amphetamine is not simply to take the role of norepinephrine as a transmitter. It appears that amphetamine molecules can mimic norepinephrine molecules so closely that the former take the place of the latter in storage sites inside the presynaptic neuron. This results in norepinephrine molecules being displaced from storage and entering the space between the presynaptic and postsynaptic neurons. Because of their increased concentration there, a larger number attach to protein molecules in the wall of the postsynaptic neuron, change the shape of the protein molecules, and trigger a signal. Hence, the activity of the neurons is enhanced, and sensations of euphoria and increased alertness are experienced.

The amphetamine molecule can exist in two forms, one being the mirror image of the other [like lactic acid (33) and other *chiral molecules]. The form that rotates plane polarized light to the right is called *dexedrine* and is much more potent than its mirror-image molecule. The two differ in shape, like hands, and only one of them readily fits a specific protein "glove."

The *methamphetamine* molecule differs from that of amphetamine in having a —CH₃ group in place of one of the hydrogen atoms of the —NH₂ group. This stimulating molecule is sold as *Methedrine* and more commonly known as *speed*.

CAFFEINE (152) $C_8H_{10}O_2N_4$

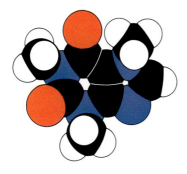

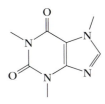

Caffeine is the component of coffee and tea that stimulates the cerebral cortex; it does so by inhibiting an enzyme *(phosphodiesterase)* that in turn inactivates a certain form (called "cyclic AMP") of the energy-supply molecule ATP (98). That caffeine can pass itself off as ATP should not be too surprising: A comparison of their structures shows that both include two fused heterocyclic rings (rings containing atoms of other elements as well as of carbon).

A typical cup of coffee or tea contains about a tenth of a gram of caffeine; coffee is obtained from the roasted seeds of *Coffea arabica,* and tea from the fermented leaves of *Camellia thea.* Caffeine also occurs in the seeds of the West African kola plants *(Cola acuminata* and *C. nitida),* which are now grown extensively in South America. Extracts from these plants are used to flavor (and add stimulant action to) cola drinks, in place of the cocaine they originally contained. [Cocaine, which is also obtained from the seed of *C. nitida,* acts like amphetamine (151).]

Theobromine is a close relative of caffeine; its molecule differs from that of caffeine only in the replacement of an —N—CH_3 group by —N—H. It is the stimulant in chocolate (127).

The seeds of the kola plant *(Cola nitida).*

TETRAHYDROCANNABINOL   (153) $C_{21}H_{30}O_2$

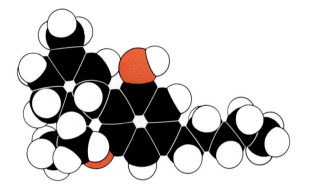

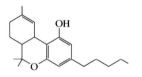

Tetrahydrocannabinol is the active component of *cannabis*, which is obtained from *Cannabis sativa*, a plant in the same family as the hop plant used in beermaking (193). *C. sativa* has been grown widely for its fibers (called *hemp*); canvas takes its name from cannabis, for it was woven of fibers obtained from the plant.

The plant exudes a resin that covers its flowers and nearby leaves, and this resin is incorporated in some preparations that humans take to consume cannabis. *Marijuana* denotes the dried, crushed leaves and flowers of the plant; *hashish* is more precisely the resin. The mode of action of this structurally highly complicated molecule has not yet been determined.

The cannabis plant *(Cannabis sativa)*, the source of hemp for canvas and of cannabis.

NASTY COMPOUNDS

To many people a "chemical" is something to be avoided, an evil and artificial corruption of nature. The preceding pages should have dispelled that notion, for they have shown that *everything tangible* is a chemical. All that is good is chemical, just as is all that is bad. All that is natural is chemical, just as is all that is artificial. Nevertheless, it cannot be denied that some chemicals have done evil—sometimes by design and sometimes by accident. Evil chemicals, like evil deeds, capture bigger headlines than the good. Chemicals that poison, which for some are the only public manifestation of chemistry and chemicals, also often poison the image of chemistry in the public's eye.

This section describes a ragbag of molecules that have hit the headlines on various occasions by virtue of their unpleasant effects on people and other animals—which include poisoning them, deforming them, or simply tearing them apart. I found it rather depressing to write about them, and hope that you will remember instead the better lives of other molecules.

. . . .

TRINITROTOLUENE   (154) $C_7H_5O_6N_3$

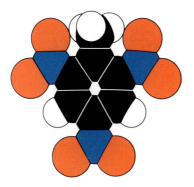

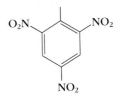

The trinitrotoluene molecule is a toluene molecule (24) to which have been attached three *nitro* groups (—NO_2).

Trinitrotoluene is the high explosive *TNT.* It is explosively unstable because it is an assembly of carbon atoms that are on the brink of *oxidation: Its oxygen atoms have only to be nudged into slightly different locations for them to be able to swoop down on the carbon and hydrogen atoms of the benzenelike ring and carry them off as carbon dioxide (4) and water (6), leaving the nitrogen atoms to fall together and move off as gaseous nitrogen (3). In an instant, therefore, the compact molecule can be converted into a voluminous cloud of gas, and the pressure wave of its expansion is the destructive shock of the explosion. That is the general function of high explosives: the sudden creation of a large volume of gas from a small volume of liquid or solid. Typically, 1 gram of explosive suddenly produces about 1 liter of gas, corresponding to a thousandfold increase in volume. There is not enough oxygen in a TNT molecule to oxidize it completely so that its explosion is marked by a good deal of black smoke.

The rearrangement of the TNT molecule is achieved by hammering it with a sharp pressure wave from another and more easily induced explosion in a detonator. A typical detonator is the solid *lead azide* [Pb(N$_3$)$_2$], which contains the azide ion N$_3^-$. It is more sensitive to shock than TNT; the azide ion shakes itself apart into nitrogen gas when lead azide is struck or exposed to the shock of an electrical discharge.

One advantage of TNT, to munitions manufacturers if not to humanity as a whole, is that it melts at a low temperature (80°C). It can thus be poured into shells and bombs.

NITROGLYCERIN   (155)   C$_3$H$_5$O$_9$N$_3$

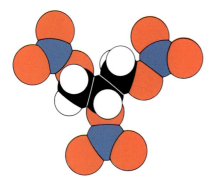

The nitroglycerin molecule is derived from the glycerol (alias glycerin) molecule (34) by replacing the hydrogen atom of each —OH group with a nitro group (—NO$_2$).

Nitroglycerin is an oily colorless liquid. In it the nitroglycerin molecules contain all the seeds of their own destruction, for the carbon and hydrogen atoms can be converted to carbon dioxide and gaseous water, and the nitrogen atoms can pair up without any input of oxygen from outside. A mechanical shock will so distort the nitroglycerin molecule that its atoms can change partners and, as with TNT (154), generate a rapidly expanding cloud of gas.

Nitroglycerin is very unstable and very sensitive to shock, impact, and friction, as may be familiar from the film *The Wages of Fear*. It is so unstable that it is normally dissolved in an absorbent material, a process that gives *dynamite*. (This invention, together with the oilfields in Russia that he owned, provided Alfred Nobel with his fortune.) Originally the absorbant material was *kieselguhr*, a clay, but modern dynamites use a mixture of wood flour, ammonium nitrate, sulfur, and sodium nitrate.

BIS(2-CHLOROETHYL)THIOETHER   (156) $C_4H_8SCl_2$

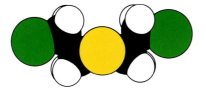

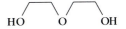

This sinister little molecule is *mustard gas,* used first at Ypres in 1917 and then stockpiled in large quantities during World War II. Mustard gas, which is in fact a volatile liquid, is odorless and therefore is not immediately detectable by smell. Where it touches the skin and is inhaled, it forms blisters. Those who do not die at once, or from the infections that follow the blistering, suffer a generalized poisoning that renders them ill for the rest of their lives. "Improvements" to the molecule that reflect the progress of civilization since Ypres include the lengthening of the hydrocarbon chain so that the molecule can pass itself off as a hydrocarbon and more easily penetrate protective rubber clothing.

DIETHYLENEGLYCOL   (157) $C_4H_{10}O_3$

The great Austrian wine scandal of 1985 concerned the contamination of large quantities of that country's sweet wines with diethyleneglycol in an attempt to modify the taste, making it resemble that of more expensive wines. Had the added substance been ethylene glycol (65), that would have been bad enough: Ethylene glycol is the substance added to automobile cooling systems to lower the freezing point of the coolant. It is sweet but causes damage to the liver. However, the criminals, obviously not knowing enough about molecules, seem to have made a double error: They confused glycerol (34), which is present in many sweet wines and would have been acceptable, with ethylene glycol, and they ignored the prefix di-; what they added was a substance that is used to some extent as an antifreeze but is actually an industrial solvent.

THALIDOMIDE   (158) $C_{13}H_{10}O_4N_2$

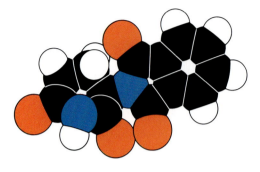

Thalidomide (which was marketed under many different names) became a popular sedative and mild hypnotic when it was first introduced to the market in 1956; even massive overdoses were not lethal. In the early 1960s, however, an increasing number of congenitally deformed (teratogenic) babies were born to mothers who had taken the drug during the first three months of pregnancy, and the drug was soon withdrawn from sale in Europe. (It had never been sold in the United States, because of certain neurological side effects that had been reported.) Since then, tests for the teratogenicity of potential drugs have become mandatory.

The chemical role of thalidomide may be related to its ability to combine with some amines, related to putrescine (131), that play a role in DNA replication. Recall (page 89) that some compounds are *chiral and exist in left-handed and right-handed forms. This is far from being an academic point, for it is now known that only the right-handed form of thalidomide is teratogenic. The commercial product was a mixture of left and right; all might have been well if only left-handed molecules had been produced and ingested.

SEX

After the horrific activities of the molecules encountered in the previous section, it is a relief to come to molecules with a more agreeable—indeed glorious—purpose.

An austere view of the ultimate purpose of sex is that it is a way of ensuring that a particular fragile pattern of hydrogen bonds (page 11) survives in the world. By this I have in mind the DNA molecule that inhabits the nucleus of every cell and propagates genetic information from generation to generation. That molecule, the famous *double helix,* is held in shape largely by the hydrogen bonds between its components, and its replication depends on the sequence of components acting as a template for the construction of a copy. It acts as a template through the hydrogen bonds its components can form, so that reproduction is a replication of hydrogen-bonding patterns. From this point of view, evolution is the consequence of the competition between patterns of hydrogen bonds.

The unconscious strategy of these patterns is to encourage their own replication by a variety of different devices. These devices are controlled by the hormones described in this section.

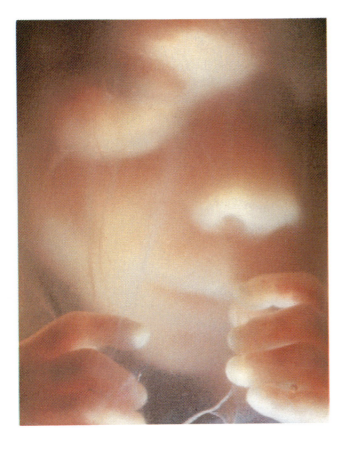

A human fetus at about 5½ months.

TESTOSTERONE   (159) $C_{19}H_{28}O_2$

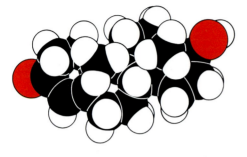

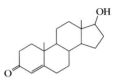

The testosterone molecule is a shortened version of the cholesterol molecule (38): The latter's ring structure is preserved, but the hydrocarbon tail has been lost. In addition, the —OH group of the cholesterol molecule has been replaced by a doubly bonded oxygen atom, so that testosterone is a *ketone; hence the ending -one. Testosterone is the male sex hormone. Its secretion from the *cells of Leydig* in the testes is initiated at puberty and controls the development of the secondary sexual characteristics, including differences in the skeleton, the voice, the pattern of body hair, patterns of behavior, and the organs of reproduction themselves. Testosterone molecules also induce the retention of nitrogen and hence encourage protein formation *(anabolism)*, which leads to enhanced musculature.

Testosterone is a member of the class of compounds called *steroids*, of which cholesterol is both a member and a metabolic precursor. Related anabolic steroids have been used to encourage the growth of muscles in athletes, but some have the awkward side effect—the term seems inappropriate—of causing a persistent erection.

ESTRADIOL   (160) $C_{18}H_{24}O_2$

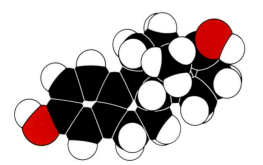

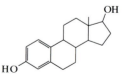

The estradiol molecule has the same ring structure as testosterone but carries an —OH group on the benzenelike ring and lacks a —CH$_3$ group at the junction of that ring and its neighbor.

It is extraordinary what difference an extra —CH$_3$ group and the slight rearrangement of a ring of atoms can make: Estradiol is one of the principal female sex hormones. It is released at puberty, maintains the secondary sexual characteristics, and then decreases in abundance at menopause. The tissues on which estradiol acts bind it strongly; it stimulates the synthesis of the RNA molecules that mediate the interpretation of DNA, and growth occurs accordingly.

The *ethynodiol diacetate* molecule differs from the estradiol molecule in being an *ester; in it, each of the two —OH groups of the hormone molecule has been combined with an acetic acid molecule (32). In addition, it carries a triply bonded pair of carbon atoms.

Oral contraceptives, which are arguably one of chemistry's great contributions to liberty, are often a combination of an *estrogen* and a *progestogen;* they act by maintaining the high levels of hormones that are characteristic of pregnancy, when ovulation is suppressed. Ethynodiol diacetate [*ethyne* is the modern name for *acetylene* (HC≡CH), and the molecule contains this group] is an example of a progestogen. Progestogen-only preparations are available as "mini-pills," so called because they are used in smaller doses. They act by thickening the mucus at the neck of the womb, so that it forms a barrier against sperm, and they modify the lining of the womb so that it is unfavorable to the growth of an egg.

Venus and Mars, 1480s(?), by Alessandro di Mariano Filipepi (called Sandro Botticelli).

GLOSSARY

An asterisk denotes another entry in this glossary.

acid   A substance that can donate a hydrogen ion (a proton) to another. Examples include sulfuric acid (10) and nitric acid (13).

alcohol   An organic compound containing the group —OH attached to a carbon atom that is not part of an aromatic ring or carries another oxygen atom. Ethanol (27) is an example.

aldehyde   An organic compound containing the group

which is normally abbreviated to —CHO. Examples include acetaldehyde (30).

alkali   A water-soluble *base. One example is sodium hydroxide (NaOH); another is ammonia (7).

alkaloid   A naturally occurring organic compound containing nitrogen that acts as a base. Many are poisonous. An example is caffeine (152).

amide group   The group

$$\overset{\displaystyle O}{\underset{\displaystyle \parallel}{}}$$
—C—NH₂

See *peptide.

amino group   The group —NH₂. Substances that contain this group (including those in which the hydrogen atoms have been replaced by other hydrocarbon groups) are called *amines*. Examples are hexamethylenediamine (71) and trimethylamine (132).

amino acid   A substance with molecules that contain both an *amino group and a *carboxyl group. An example is glycine (73).

anion   A negatively charged *ion.

antioxidant   A substance that decreases the rate at which another substance is *oxidized. An example is BHA (42).

aromatic compound   A substance containing one or more benzene rings (23). Aromatic compounds, though *unsaturated, do not undergo the normal reactions of unsaturated compounds. Their special stability is related to the ability of the bonding electrons to spread around the planar ring.

atactic   Lacking regularity. A polymer is atactic if the groups attached to the backbone are not arranged in any regular geometric pattern. See *isotactic.

atom   The smallest part of an element that retains its properties. An atom consists of a minute positively charged nucleus surrounded by a cloud of *electrons.

base   A substance that can accept a hydrogen ion from an *acid. An example is ammonia (7), which accepts H^+ and forms the *ammonium ion* NH_4^+.

bond   A chemical link between two atoms. In an *ionic* bond, the attraction is between the opposite charges of two neighboring *ions. In a *covalent* bond, the two atoms share a pair of electrons that lie between them.

carbonyl group   The group $\diagdown C = O$. When a hydrogen atom is attached to the carbon atom, the resulting compound is called an *aldehyde. When only carbon atoms are attached, the resulting compound is called a *ketone. See also *carboxyl group and *amide group.

carboxyl group   The group

a *carbonyl group next to a *hydroxyl group. The carboxyl group is normally abbreviated to —COOH. See *carboxylic acid.

carboxylic acid   An organic compound containing the carboxyl group. The hydroxylic hydrogen can be lost as a *hydrogen ion, so that substances with this group are *acidic. An example is formic acid (31).

catalyst   A substance that facilitates a reaction without itself being consumed. (The Chinese term for catalyst, "marriage broker," conveys the sense.)

cation   A positively charged *ion.

chain reaction   A reaction in which the product of one step is a reactant in a later step, which produces a reactant for another step, and so on.

chiral   Handed. A chiral molecule is one that is distinguishable from its mirror image. An example is lactic acid (33).

cortex   The outer, most highly developed, and evolutionally most recent part of the brain.

crystal   A solid in which the atoms, ions, or molecules lie in an orderly and virtually endlessly repeated arrangement.

distillation   Separation of the components of a mixture on the basis of their different volatilities.

electron   An elementary particle with a negative charge. In an *atom, the electrons are arranged in shells around the nucleus, and only those in the outermost shell take part in the formation of chemical *bonds.

electron pair   Two electrons responsible for a chemical *bond. See also *lone pair.

electronegative atom   An atom that draws electrons toward itself in a molecule (the most electronegative atoms are fluorine, chlorine, oxygen, and nitrogen).

emulsion   A dispersal of one liquid as minute particles (each one consisting of many molecules) in another liquid. An example is milk (page 57).

enzyme   A substance that facilitates a biochemical reaction; a biological *catalyst.

epithelium   A type of tissue that covers a body or a structure of the body.

ester   The outcome of the reaction between a *carboxylic acid and an *alcohol. An example is tristearin (36).

fatty acid   A *carboxylic acid, especially one with a long hydrocarbon chain.

fluorescence   The emission of longer-*wavelength light immediately following the absorption of shorter-wavelength radiation. *Phosphorescence* is similar, but may persist after the stimulating radiation has been extinguished.

ganglion An aggregation of nerve-cell bodies outside the central nervous system.

glycoside A sugar molecule in which the hydrogen atom of an —OH group is replaced by another group. A *glucoside* is a glycoside in which the sugar is glucose (79).

heterocyclic molecule A molecule containing at least one ring of carbon atoms, with at least one other type of atom (normally nitrogen or oxygen) in the ring. An example is 2,6-dimethylpyrazine (127).

hydrocarbon A compound of carbon and hydrogen alone. The simplest example is methane (16).

hydrogen bond A link formed by a hydrogen atom between two *electronegative' atoms, and denoted A···H—B. See page 12.

hydrogen ion The ion left when a hydrogen atom loses its only electron; a bare hydrogen nucleus, a proton. In water, a hydrogen ion would become attached to a water molecule to form the unit H_3O^+.

incandescence The emission of light by a hot substance. All wavelengths of radiation are present but the greatest intensity shifts from red to blue as the temperature is raised.

infrared radiation Electromagnetic radiation whose *wavelength is longer than that of visible red light. Infrared radiation is responsible for the transmission of radiant heat.

ion An electrically charged atom or group of atoms. If an atom or molecule loses one or more electrons, it becomes a positively charged *cation.* If it gains one or more electrons, it becomes a negatively charged *anion.*

isomer A molecule built from the same atoms as another but in a different arrangement.

isotactic Geometrically regular. A polymer is isotactic if all the groups attached to the backbone are arranged in the same geometrical pattern. See *atactic.

ketone An organic compound containing the *carbonyl group $\diagdown C{=}O$, to which other carbon atoms are attached. An example is testosterone (159).

limbic system A network of neurons encircling the brain stem, thought to control the emotions and translate them into actions. (*Limbus* is Latin for "hem.")

lipid A naturally occurring substance that is soluble in organic solvents but not in water.

lone pair A pair of electrons not involved in bond formation.

magnetism The tendency of a substance to move into a magnetic field. This is more strictly called *paramagnetism.* It arises from the presence of one or more electrons that are not taking part in *bond pairs or *lone pairs. Each behaves as a tiny magnet with a magnetic field that is not canceled by the other (missing) member of a pair.

monomer A unit from which a *polymer is built.

osmotic pressure The pressure needed to prevent the flow of a solvent through a semipermeable membrane. A solvent tends to flow through such a membrane from a less to a more concentrated solution. This tendency can be opposed by exerting a pressure on the more concentrated solution. The magnitude of the pressure required to stop the flow is the osmotic pressure.

oxidation Reaction with oxygen, as in combustion. This is the original meaning of the term, and the one used in this book. It has also been generalized to cover a wide range of reactions with a similar outcome—that is, electron loss.

peptide group The group

Molecules containing this group are *peptides* and include the polypeptides (76).

phenol A substance in which a *hydroxyl group is attached directly to a benzene ring. An example is vanillin (126).

photochemical reaction A chemical reaction induced by light.

polarized light Electromagnetic radiation in which the oscillating electric fields all lie in the same plane.

polarized light microscopy The study of transparent substances, including crystals, with a microscope in which the sample lies between two polarizing filters.

polymer A molecule formed by connecting smaller molecules together to form a string or network.

prokaryote A cell without a membrane-bound nucleus.

protein A molecule formed by stringing together *amino acid molecules. Proteins, which include *enzymes, are polypeptides (76) that carry out many of the biochemical processes of living things.

radical A fragment of a molecule containing at least one electron not present as part of a bond pair or a *lone pair.

reduction The addition of hydrogen and the removal of oxygen; the opposite of *oxidation. (This original meaning has been generalized to include the addition of electrons to a substance by any means.)

salt The ionic product of a reaction between an *acid and a *base.

saturated compound An organic compound that does not contain carbon-carbon multiple bonds. An example is ethane (17), as distinct from ethylene (47).

surfactant A surface-active agent; that is, one that accumulates at the interface between two liquids and modifies their surface properties. An example is the stearate ion (43).

synapse The junction between two neurons, or nerve cells.

thalamus A part of the vertebrate brain just behind the cerebrum; an important entry point into the brain from the rest of the nervous system.

transmutation The conversion of one element into another by a process taking place in the nucleus.

ultraviolet light Electromagnetic radiation whose *wavelength is shorter than that of visible violet light.

unsaturated compound An organic compound that contains carbon-carbon multiple bonds. An example is ethylene (47), as distinct from ethane (17).

wavelength The distance between neighboring peaks of a wave of electromagnetic radiation (or any other periodic wave).

FURTHER READING

Atkins, P. W. *General Chemistry*. New York: Scientific American, in press.

Austin, G. T. *Shreeve's Chemical Process Industries*. New York: McGraw-Hill, 1984.

Billmeyer, F. W. *Textbook of Polymer Science*. New York: Wiley-Interscience, 1984.

Bowman, W. C., and M. J. Rand. *Textbook of Pharmacology*. Oxford, England: Blackwell, 1980.

Brady, G. S., and H. R. Clauser. *Materials Handbook*. New York: McGraw-Hill, 1977.

Considine, D. M., and G. Considine. *Encyclopedia of Chemistry*. New York: Van Nostrand Reinhold, 1984.

Corbin, A. *The Foul and the Fragrant*. Cambridge, Mass.: Harvard University Press, 1986.

Coultate, T. P. *Food: The Chemistry of Its Components*. London: Royal Society of Chemistry, 1984.

Coyle, J. D., R. R. Hill, and D. R. Roberts. *Light, Chemical Change, and Life*. Philadelphia: Open University Press, 1982.

Crossland, J. *Lewis's Pharmacology* London: Churchill Livingstone, 1980.

Curtis, H. *Biology*. New York: Worth, 1983.

Grayson, M. *Concise Encyclopedia of Chemical Technology*. New York: Wiley, 1985.

Greenwood, N. N., and A. Earnshaw. *Chemistry of the Elements*. Oxford, England: Pergamon, 1984.

Hanssen, M., with Jill Marsden. *E for Additives*. Wellingborough, England: Thorsons, 1984.

Kent, J. A. *Riegel's Handbook of Industrial Chemistry*. New York: Van Nostrand Reinhold, 1974.

McGee, H. *On Food and Cooking*. New York: Scribner's, 1984.

Medved, E. *Food: Preparation and Theory*. Englewood Cliffs, N. J.: Prentice-Hall, 1986.

Nassau, K. *The Physics and Chemistry of Color*. New York: Wiley-Interscience, 1983.

Nogrady, T. *Medicinal Chemistry*. New York: Oxford University Press, 1985.

Parker, S. P. *Encyclopedia of Chemistry*. New York: McGraw-Hill, 1982.

Press, F., and R. Siever. *Earth*. 4th ed. New York: Freeman, 1986.

Raven, P. H., R. F. Evert, and S. E. Eichhorn. *Biology of Plants*. New York: Worth, 1983.

Selinger, B. *Chemistry in the Market Place*. London: Murray, 1978.

Shapiro, R. *Origins*. Portsmouth, N. H.: Heinemann, 1983.

Snyder, S. H. *Drugs and the Brain*. New York: Scientific American, 1986.

Stryer, L. *Biochemistry*. 3d ed. New York: Freeman, 1988.

Zapsalis, C., and R. A. Beck. *Food Chemistry and Nutritional Biochemistry*. New York: Wiley, 1985.

SOURCES OF THE ILLUSTRATIONS

Line illustrations by Tom Cardamone Associates, Inc.

• • •

page 1   Joan Miró, *The Beautiful Bird Revealing the Unknown to a Pair of Lovers,* 1941, gouache and oil wash, 18 × 15 in (45.7 × 38.1 cm), collection, The Museum of Modern Art, New York, acquired through the Lillie P. Bliss Bequest

page 2   C. F. Quate and Sang-il Park/Stanford University

page 3   Heather Angel/Biofotos

page 4 (top)   Chip Clark

page 4 (bottom)   C. F. Quate and Sang-il Park/Stanford University

page 12   Travis Amos

• • •

page 13   David Whillas/CSIRO Atmospheric Research, Australia

page 14   Tui De Roy

page 15   Sceptre Books, London—

adapted from *The Cambridge Encyclopedia of the Earth Sciences.*

page 16   Travis Amos

page 18 (top)   Thomas Hovland/Grant Heilman

page 18 (bottom)   Chip Clark

page 20   Dr. J. Burgess/Science Photo Library, Photo Researchers

page 21   Susan Middleton

page 24   Chip Clark

page 25   Tom Bean

page 28   Vulcain Explorer/Science Source, Photo Researchers

page 29   Charles Arneson

• • •

page 33   Susan Schwartzenberg/Exploratorium

page 35   Pacific Gas and Electric

page 37   Manfred Kage/Peter Arnold, Inc.

page 45   Travis Amos

page 46   Martin Miller/University of California, Davis

page 47 (bottom left)   Edward Ross

page 47 (bottom right)   Heather Angel/Biofotos

page 49   Sondra Barrett

page 51   Susan Middleton

page 52   L. Egede-Nissen/Biological Photo Service

page 54   Tony Morrison/South American Pictures

page 59   Heather Angel/Biofotos

page 60   D. Cavagnaro/Peter Arnold, Inc.

page 61   Science Photo Library/Photo Researchers

• • •

page 65   Susan Middleton

page 66   Left to right: Arthur Aykanian, *Spoonstraw,* 1968, plastic, gift of Winkler/Flexible Products, Inc.; Earl S. Tupper, *Drinking Glasses,*

1954, plastic, gift of Tupper Corporation; Gino Colombini, *Pail*, 1954, plastic, 10.5 in. high, Philip Johnson Fund; Oscar Kogoj, *Aspirator*, 1974, rubber and plastic, gift of Ciciban Shoe and Children's Ware Factory; Gene Hurwitt, *Containers*, 1966, colored plastic, purchase; all in the collection, The Museum of Modern Art, New York

page 67   Andrew Holik/General Electric Corporate Research and Development, Schenectady, New York

page 68   Christopher Springmann/ The Stock Market of New York

page 70   Ted Horowitz/The Stock Market of New York

page 74   Travis Amos

page 75   Andy Warhol, *Self-Portrait*, 1966, synthetic polymer paint and enamel silk-screened on six canvases, each 22⅝ in. square, The Museum of Modern Art, New York (The Sidney and Harriet Janis Collection)

page 77   Jacques Jangoux/Peter Arnold, Inc.

page 82   National Institutes of Health/Science Source, Photo Researchers

page 86   Fibers Division, Monsanto Chemical Company

page 87 (top)   Fibers Division, Monsanto Chemical Company

page 87 (bottom)   Science Photo Library/Photo Researchers

page 91   Adapted from *The Structure and Action of Proteins*, Richard E. Dickerson and Irving Geis, W. A. Benjamin, Inc., 1969

page 92   Thea Gabra-Sanders/Teaching and Research Centre, Western General Hospital, Edinburgh

page 93   John Shaw

page 94 (top)   Mark Denny/Hopkins Marine Station

page 94 (bottom)   Adapted from *The Structure and Actions of Proteins*, Richard E. Dickerson and Irving Geis, W. A. Benjamin, Inc., 1969

page 97   Mia Tegner/Scripps Institute of Oceanography

page 98   Philip Rosenberg

page 100   Pasteur Institute

page 102   Institute of Paper Chemistry

page 103   Charles Arneson

page 104   Ken Hoy/Ardea
. . .
page 105   Sondra Barrett

page 106 (left)   From *Tissues and Organs: A Text-Atlas of Scanning Electron Microscopy*, Richard Kessel and Randy Kardon, W. H. Freeman and Company, 1979

page 111   J. L. Mason/Ardea

page 113   Runk/Schoenberger/Grant Heilman

page 115   Travis Amos

page 117   From *Tissues and Organs: A Text-Atlas of Scanning Electron Microscopy*, Richard Kessel and Randy Kardon, W. H. Freeman and Company, 1979

page 118   Travis Amos

page 121   Travis Amos

page 122   Carl Wirsen/Woods Hole Institute

page 124   From *Tissues and Organs: A Text-Atlas of Scanning Electron Microscopy*, Richard Kessel and Randy Kardon, W. H. Freeman and Company, 1979

page 129   Travis Amos

page 130   Travis Amos

page 131   Travis Amos

page 135   J. L. Mason/Ardea

page 139   J. L. Mason/Ardea

page 141   Peter Steyn/Ardea

page 142   Susan Middleton

page 143   Peter Arnold/Peter Arnold, Inc.
. . .
page 145   Michel Viard/Peter Arnold, Inc.

page 146   From *Tissues and Organs: A Text-Atlas of Scanning Electron Microscopy*, Richard Kessel and Randy Kardon, W. H. Freeman and Company, 1979

page 152 (top)   Travis Amos

page 152 (bottom)   John Heywood/ University of Miami

page 153   William Sheridan/University of North Dakota

page 154   Travis Amos

page 155   Heather Angel/Biofotos

page 157   John Shaw

page 159   P. Morris/Ardea
. . .
page 161   W. H. Hodge/Peter Arnold, Inc.

page 165   Adapted from *Drugs and the Brain,* Solomon Snyder, Scientific American Books, Inc., 1986

page 166   J. N. A. Lott, McMaster University/Biological Photo Service

page 169   W. H. Hodge/Peter Arnold, Inc.

page 170   Kent Wood/Peter Arnold, Inc.

page 175   Petite Format/Nestle/Science Source, Photo Researchers

page 177   Alessandro di Mariano Filipepi, called Sandro Botticelli, *Venus and Mars,* 1480s(?), wood, 69.2 × 173.4 cm, The National Gallery, London

INDEX

α-amino acid, 89
α-keratin, 90
abherent coating, 72
absorption, and color, 30, 149
Acer saccharum, 98
acetaldehyde, 46
acetanilide, 163
acetic acid, 48
Acetobacter, 48
acetylcholine, 47
acetylsalicylic acid, 162
Achras sapota, 133
acid, 179. *See also* individual
   acids.
acid rain, 27, 30
Acrilan, 83
acrolein, 99, 123, 158
acrylic, 80, 83
acrylic ester polymer, 74
acrylic paint, 74
acrylonitrile, 82
actin, 117
addiction, 162
adenine, 119
adenosine triphosphate, 119
adipic acid, 84
adrenaline, 167
air, composition of, 14
   liquid, 17

alanine, 89
albumin, decomposition of,
   123
alcohol, 42, 43, 179
   dehydrating effect, 44
   dehydrogenase, 42, 46
   hangover, 46
   and intoxication, 43
   metabolism, 44, 46
   origin of name, 42
   and sedatives, 44
aldehyde, 42, 45, 179
alertness, 168
Alka-Seltzer, 163
alkali, 179
alkaloid, 110, 179
alkyd paint, 55
alkyd resin, 82
Allium cepa, 131
Allium sativum, 131
allyl propyl disulfide, 131
almonds, odor of, 126
amide group, 179
amide link, 84, 86
amino acid, 88, 179
   chirality, 89
   essential, 88
   and ethnic cookery, 88
amino group, 85, 88

ammonia, 26
   and flatulence, 140
   in garlic and onion, 131
   pungency, 26
   in urine, 143
Ammonians, 26
ammonium nitrate, 31
amorphous regions, 68
AMP, 169
amphetamine, 167
   action of, 168
amygdalin, 127
amyl dimethylaminobenzoate,
   160
amyl nitrite, 163
amylopectin, 100
amylose, 100
anabolic steroids, 176
anabolism, 176
analeptic, 137
anethole, 136
angiosperms, 110
aniline, 163
animals, color changing in,
   159
   diving, color of meat, 118
   smells of, 140
anion, 10, 180, 181
aniseed, 136

annatto, 152
anosmia, 125
ant venom, 47
Antheraea pernyi, 93
anthocyanidin, 156
anthocyanin, 156
anthrax, 45
antianxiety drugs, 164
anticline, 34
antidiuretic hormones, 44
antifreeze, 80
antimony sulfide, 42
antioxidants, 180
   in foods, 60
   in plastics, 69
   in smoke, 123
antipruritic, 137
anxiety, 164
apple, color of, 156
   ripening of, 127
   taste of, 111
apricot, color of, 152
   cyanide in, 127
aramid, 84
argon, 16
armpit, odor of, 58, 140
aromatic compound, 40
arteries, hardening of, 56
arthropod, exoskeleton of, 104

187

aspartame, 109
aspartic acid, 109
Aspergillus oryzae, 96, 113
aspirin, 163
astaxanthin, 154
atactic molecules, 73, 180
atherosclerosis, 56
atmosphere, composition of, 14
    of Mars, 21
    origin of, 15
    of Venus, 21
atom, 3, 180
    color code for, 5
ATP, 119, 147, 169
Austrian wine scandal, 173
autonomic nervous system, 168
axillary region, odor of, 140
azide ion, 172
azodicarbonamide, 73
Azotobacter, 19

β-keratin, 93
Bacillus acidi levolactii, 49
Bacillus licheniformis, 62
Bacillus subtilis, 62
bacteria, purple sulfur, 123.
    See also specific bacteria.
bactericide, 21
bacteriorhodopsin, 147
baking, 22
baking powder, action of, 22
    and tartaric acid, 22
barbecue, 117
    odor of, 123
barley malt, 44
base, 180
    organic, 119
basidiocarp, 121
bay, oil of, 135
beans, indigestion and flatus,
    99
bee, vision of, 155
bee's purple, 155
beef fat, 54
    color of, 54
    taste of, 120
    tenderness of, 54
beer, 44, 113
    color of, 158

beet, color of, 156
    sugar, 98
    taste of, 84
Beggiatoa, 122
belch, 134
benzaldehyde, 126
benzedrine, 168
benzene, 39
    formula of, 8
benzenebergs, 24
benzodazepines, action of, 44
benzol, 40
benzyl acetate, 133
Bertholletia excelsa, 88
Beta vulgaris, 84, 98
betacyanin, 156
BHA, 60
bicarbonate, 22
bile, 143
bile compound, 56
bilirubin, 143
biodegradable, 63
bis (2-chloroethyl) thioether,
    173
bitter almonds,
    odor of, 126
bitterness, 106, 110
bittersweet, 110
Bixa orellana, 152
bixin, 152
blackberries, color of, 156
blast furnace, 40
bleach, 31
bleaching of hair, 92
blood, oxygen transport in, 90
    plasma and cholesterol in,
        56
blood-brain barrier, 168
blood oranges, 12
blood sugar, 96
blue cheese, odor of, 128
Bombyx mori, 93
bond, 5, 180
    characteristic number of, 7
    double, 6
    hydrogen, 12
    ionic, 180
    triple, 6
Botrytis cenerea, 50

bougainvillea, color of, 156
brain stem, 164, 167
branched chain, 38
Brassica napus, 54
Brassicaceae, 59
Brazil nuts, 88
bread, crusty, aroma of, 130
    stale, 101
breadmaking, 22
    and cereal starches, 101
Brevibacterium, 121
brewer's yeast, 101
Brie, 26
brighteners, 62
brittleness, 12
bromine, 4
brown, 158
brown crust, 158
brown sugar, 97
bruise, 158
    and nonbrowning fruits, 159
Brussels sprouts, and
    flatulence, 140
BTX gasoline, 41
bulletproof vests, 87
butane, 36
butanedione, 58
butanoic acid, 57, 102
    in chocolate, 139
butter, 57
    color of, 57, 151
    flavor of, 58
butyl rubber, 79

cabbages, and flatulence, 140
cadaverine, 144
caffeine, 110, 169
calcium ions, and
    atherosclerosis, 56
camel's hump, 54
camelia, 156
Camellia thea, 169
Camembert, 26
camphor, 136
    as antipruritic, 137
camping gas, 36
candy, 97
    role of glucose in, 96
cannabis, 113, 170

Cannabis sativa, 170
canthaxanthin, 154
canvas, 170
capsaicin, 115
Capsicum annum, 115
Capsicum frutescens, 115
caramel, 99
    color of, 158
    odor of, 123
carapaces, color of, 154
caraway, 134
carbohydrates, 95
    photosynthesis of, 150
carbon, 8
carbon dioxide, 21
    in beer, 113
    and carbonated drinks, 22
    and greenhouse effect, 22
    and maple syrup harvest, 98
carbon monoxide, 35
    poisoning, 90
carbonate rock, 21
carbonated drinks, 22
carbonic acid, 21
carbonyl group, 8, 58, 180
carboxyl group, 180
carboxylic acid, 42, 180
Carthamus tinctoris, 55
carminative action, 134
carotene, 151
    in fat, 54
    in margarine, 59
carotenoids, 149
carpet fibers, 87
carpets, 83
carrots, color of, 151
carvone, 133
cassette tape, 81
catalase, 45
catalyst, 180
catalytic converter, 39
    reforming, 38
cataracts, removal of, 147
caterpillar, stinging, 47
cation, 10, 180, 181
cats, sense of taste in, 106
cattle, and St John's wort, 155
cellophane, 103
cells of Leydig, 176

cellulose, 2, 95, 102
    acetate, 103
    fibers, 102
Centaurea cyanus, 156
cereals, amino acid content of, 88
cerebral cortex, 168
cetane, 37
    number, 38
chain reaction, 39, 55, 180
    and fire, 83
chalk, 21
chameleon, 159
champagne, 21
cheese, odor of, 58
chemical symbol, 6
chemoreception, 106
cherries, color of, 156
    odor of, 127
chewing gum, 79, 133
chicle, 133
Chile saltpeter, 31
chili, active ingredient in, 115
chiral, 49, 180
chlordiazepoxide, 164
chloride ion channel, 44
chlorine, 5
chlorobutadiene, 79
chlorophyll, 150
    in white wine, 157
chloroprene, 79
chocolate, 54
    aroma of, 130
    brittleness of, 55
    composition of, 138
    cool taste of, 55
    sharp melting point of, 55, 139
    soft-centered, 97
    stimulant in, 169
    taste of, 139
cholecalciferol, 100
cholesterol, 56
    and hormones, 176
    in milk, 57
    in musk, 141
chrysanthemum, color of, 157
cinchona tree, 112
Cinnamomum caphora, 137

Cinnamomum zeylanicum, 134
cinnamon, 134
cinnemaldehyde, 134
cis and trans, 10
    in rubber, 78
11-*cis*-retinal, 147
citric acid, 111, 163
citronella, 136
citrus fruit, nonbrowning bruising, 159
    taste of, 111
civet, 141
civetone, 141
claret, and gout, 49
claws, 92
cling wrapping, 71
Clostridium, 19
clove, 60, 128
    oil of, 128, 135
coagulation, of milk, 57
coal, 40
    structure of, 40
cocaine, 169
cocoa, 139
    beans and color, 158
    pod, 59
cocoa butter, 54
    sharp melting point of, 55
coconut oil, in soap, 62
cod liver oil, 55
codeine, 166
Coffea arabica, 130, 169
Coffea canephora, 130
coffee, aroma of, 130
    beans and flowers, 161
    roasting and color, 158
    stimulant in, 169
coke, 40
cola drinks, caffeine in, 169
    sourness in, 110
Cola acuminata, 169
Cola nitida, 169
collagen, 108
    and sponges, 117
color(s), 146
    absorption, 30
    by absorption, 149
    and acidity, 3
    of apples, 156

of apricots, 152
of autumn, 150, 151, 155, 157
of beef fat, 54
of beer, 158
of beets, 156
of blackberries, 156
of bougainvillea, 156
of bread crust, 158
of butter, 57, 151
of cabbage, red, 156
of carrots, 151
of chameleon, 159
of cheese, red, 152
of cherries, 156
of chrysanthemum, 157
of coffee, roasted, 158
complementary, 149
of corn, 153
of cornflower, 156
of cream, 151
of dried fruit, 27
of egg, cooked, 123
of fat, 151
of feces, 143
of flamingo, 154
of flesh, diving animal, 118
    fish, 118
of flowers, 149
of fruit juice, 27
of game and poultry, 118
of gas flame, 35
of geranium, 156
of grapes, 156
of grass, 150
of hair, 92, 159
    and gray restoration, 92
of ham, 118
of leaves, 149, 150
of lobster, 154
of mangos, 151, 153
of maple syrup, 99, 158
of margarine, 57, 59
of meat, cooked, 118
    red and white, 118
of methanol flame, 42
of milk, 151
of newsprint, 102
of orange juice, 153

origin of, 17
of paper pulp, 102
of peony, 156
of persimmon, 151, 152
of pigs, 18
of polystyrene, aged, 73
of poppy, red, 156
of raspberries, 156
of Red Sea, 20
of rhubarb, 156
of salmon, 154
of shrimp, 154
of skin, 159
of smog, 30
of strawberries, 156
of tea, 159
of tomatoes, 152
of Turacos, 143
of urine, 143
of vegetables, cooked, 150
of water, 24
of wine, 156
of yolk, 153
color wheel, 149
complementary color, 149
compound(s), 5
    aromatic, 40
    inorganic, 6
    ionic, 10
    organic, 6
    organometallic, 39
    phenolic, as antioxidants, 60
conditioner, 92
cones, in eye, 146
connective tissue, 117
contraceptives, oral, 177
cooking, and color of vegetables, 150
    ethnic, 88
cool taste, of chocolate, 55
    of menthol, 116
copolymer, 66
copper, 4
corn, color of, 153
cornflower, and acidity, 156
    color of, 3
corn oil, 55, 59
corn sugar, 96
corn syrup, 96

cortex, 43, 180
cottonseed oil, 55, 59
covalent bond, 180
coyote, odor of, 144
crab, exoskeleton of, 104
cracking, 67
cramp, 49, 101
cream, 57
   color of, 151
crease-resistance, 81
credit cards, 70
Crimplene, 81
crocetin, 152
Crocus sativus, 145, 152
crust, brown color of, 158
crystal, 180
crystalline regions, 68
crystallinity, 69
curdling, of milk, 57
cyanide group, 76
cyanidin, 156
cyanobacteria, 17, 34
cyclamate, 108
cyclic AMP, 169
cyclohexylamine, 109

Dacron, 81
dandelion, latex in, 77
death, changes after, 119, 120
   odor of, 122, 144
defecation, 115
Demospongiae, 117
demulcent, 50
detergent, 61
   biodegradable, 63
   cholesterol as, 56
   effect of, 61
   nonionic, 64
detonator, 172
dexedrine, 168
dextrose, 96
diacetyl, 58
diacetylmorphine, 166
diallyl disulfide, 131
diamorphine, 166
diatomic molecule, 18
diazepam, 164
dicarboxylic acid, 80, 86
diesel fuel, 37, 38

diethyleneglycol, 173
diets, low-calorie, 97
diffraction, x-ray, 1
digestion, in rabbit, 102
   in ruminants, 102
diglyceride, 52
dihydoxyacetone, 160
dimethylpyrazine, 138
dinosaurs, and the
   angiosperms, 110
diols, 80
disaccharide, 98
distillation, 180
   of coal, 40
   of wood, 41
DNA, 175
dogs, odor of, 144
domestic cleaners, odor of, 137
double bond, and effect on
   shape, 55
   in ethylene, 67
   in fats and oils, 55
   hydrogenation of, 59
double helix, 175
dough, flavor of, 101
   kneading, 101
dragonfly, 104
drugs, antianxiety, 164
drupe, 127
drying oil, 55
Dubonnet, 112
Dyer's oak, 157
dynamite, 172

eastern crocus, 152
Ebenaceae, 152
ebonite, 79
ebony, 152
egg, cooked, color of, 123
   rotten, 122
   yolk, color of, 153
elastic recovery, of nylon, 86
   of rubber, 79
elastomer, 77
electron, 180
electron microscopy, 1
electron pair, 6, 180
electronegative atom, 180
electrostatic shock, 86

element, 3
   chemical symbol of, 6
emollient, 50
emotion, and branched
   neurons, 167
   and odor, 124, 132, 140
emulsin, 127
emulsion, 180
endorphin, 114
enkephalin, 114, 165
enzyme, 88, 180
   proteolytic, 62
epinephrine, 168
epithelial cell, 106
epithelium, 181
epoxide, 31
Escherichia coli, 99, 100
   and flatulence, 140
essential oils, 132
ester, 42, 181
esterification of cellulose, 103
estradiol, 176
estrogen, 177
ethane, 36
ethanol, 43, 49
ethene, 67
ethinodiol diacetate, 177
ethyl alcohol, 43
ethyl group, 39
ethyl 2-methylbutanoate, 127
ethylene, and fruit ripening,
   67
ethylene glycol, 80, 173
Eugenia aromatica, 135
eugenol, 134
euphoria, sense of, 168
eutrophication, 62
exoskeleton(s), and chitin, 104
   of crab, 104
   of scorpion, 104
explosions, 171
eye, 146
   and color, 159
   and image-resolution, 148

fabric, crease-resistant, 81
fast muscle, 117
fast pain, 114
fat, 50

color of, 151
   polyunsaturated, 55
   pork, 55
   poultry, 55
   secondary role in food, 54
fatty acid, 50, 181
feces, color of, 143
feet, unwashed, 58
felt, 92
fennel, 136
FEP, 72
fermentation, 21
   of ammonia, 121
   anaerobic, 49
   of chocolate, 139
   top and bottom, 113
   of wine, 44
fermented pickles, 49
ferret, odor of, 142
fertilizer, 26, 29, 31
   phosphate, 119
fibroblast, 117
first and second smell, 126
fish, color of meat, 118
   muscle, and tenderness, 117
   odor of, 144
   smoked, 45
fish oil, 55
fixation, of nitrogen, 19
fixative, 141
flame, 21, 35
   resistance, 83
flamingo, color of, 154
flatulence, 134, 140
   and bean meals, 99
flatus (*see* flatulence)
flavone, 157
flavonoids, 155
flavonol, 157
flavor, of butter, 58
   of caramel, 99
   of chicken, cooked, 123
   of dough, 101
   of food, 132
   of margarine, 59
   of meat, 120
   of sherry, 46
   of wine, 138
flaxseed, 55

fleabane, 155
flour, wheat, 101
flowering plants, and the
    dinosaurs, 110
flowers, 132, 155
    colors of, 149
fluorescence, 16, 181
fluorescent dyes, as
    brighteners, 62
fluorescent lamp, 16, 73
fluorinated ethylene-propylene,
    72
foamed polystyrene, 73
food, 124
    aroma of, 130
    smoking, 45
food additives, adipic acid, 84
formaldehyde, 45
    in wood smoke, 123
formalin, 45
formic acid, 47
fossil fuel, 37
fragrance, of a rose, 136
fructose, 96
    in semen, 97
fruit, 124, 155
    bruised, 159
    dried, color of, 27
    ripening process, 67
fruit juice, color of, 27
fruit sugar, 97
fruity smell, 127
frying oil, aging of, 55
functional group, 8, 56
fungi, and chitin, 104
2-furylmethanethiol, 130
fusel oil, 46

GABA, 43, 164, 167
Gaillardia pulchella, 152
galactose, 99
game, color of meat, 118
gamma-aminobutanoic acid, 43
ganglion, 181
ganglion cells, damage to, 42
garlic, odor of, 131
gas, 10, 14
    camping, 36
    flame, 35

and flatulence, 99, 134, 140
    inert, 16
    mustard, 173
    natural, 34, 35
    noble, 16
gasoline, BTX, 41
gelatin, 117
geometrical isomerism, 10
geraniol, 135
geranium, color of, 156
    odor of, 136
gin and tonic, taste of, 112
ginger, 116
Glomerella cingulata, 115
glucophore, 107
glucose, 95
glue, 76
glutamic acid, 120
glycerol, 50
    and alkyd resins, 82
    and TNT, 172
    in wine, 173
glycine, 88
glycogen, 101
glycol, 80
glycoside, 127, 181
goat's milk, odor of, 58
Gossypium, 55
gout, 49
grapes, color of, 156
    Zinfandel, 52
grape sugar, 96
grass, artificial, 83
    color of, 150, 151
gravy thickening, 101
grease, 61
    and polyethylene staining,
    68
    and PTFE, 72
greenhouse effect, 22
guano, 31
gum chewing, 79
gutta-percha, 78

Haber process, 26
hair, 88
    bleaching of, 31
    brittleness of, 92
    cell, 91

color of, 92, 159
    conditioner for, 92
    extensibility of, 91
    fiber, 91
    gray, 92
    luster of, 92
    springiness of, 79
    structure of, 90, 91
Halobacterium halobium, 147
ham, color of, 118
hangover, 46
hard water, 63
hashish, 170
hay, odor of, 129, 152
head group, 61
headache, 163
heart disease, 56
Helianthus annuus, 55
helium, 16
hemicellulose, 102
hemoglobin, 90
hemp, 170
heptane, 38
heptanone, 127
heroin, 166
heterocyclic, 181
Hevea brasiliensis, 77
hexadecane, 37
hexamethylenediamine, 85
hexane, 96
hides, treatment of, 45
high-density polyethylene, 68
histamine, 47
homogenized milk, 57
honey, fructose content of, 97
    hydrogen peroxide content
    of, 31
hooves, 92
hop resin, 113
hormone, 56, 163
    antidiuretic, 44
    female, 177
    sex, 176
hornet, 47
humulone, 112
Humulus lupulus, 113
hyacinth, odor of, 126
hyacinthin, 126
hydrocarbons, 34, 181

hydrogen, 5
    ion, 181
hydrogen bonding, and
    cooking, 101
    in nylon, 84
hydrogen bonds, 12, 181
    in cellulose, 102
    and evolution, 175
    in glucose, 96
    and viscosity, 50
hydrogen cyanide, 126
hydrogen sulfide, 122
    and flatulence, 140
hydrogen peroxide, 31
    and hair bleaching, 92
    in honey, 31
hydrogenation, 59
hydrophilic, 61
hydrophobic, 61
hydroxyl group, 8
Hypericum, 155

ice, 23
    structure of, 24
    translucency of, 68
iceberg, 23
IMP, 120
incandescence, 35, 181
incandescent lamp, 16
incubators, 18
Indian blanket, 152
indigestion, 100
industrial intermediate, 45
infrared light, 181
inhibitory control, 43
inorganic compounds, 6
inosine monophosphate, 120
insulator, 68
intima, of artery, 56
invert sugar, 99
invertase, 99
ion, 10
    hydrogen, 181
ionic solid, 11
ionic bond, 180
ionic compound, 10
ionone, 128, 152
ischemic heart disease, 56
Isla Santa Cruz, 14

isoamyl acetate, 127
isobutylene, 79
isomer, 10, 38, 181
isomerism, geometrical, 10
isooctane, 38
isoprene, 77, 132
   and plant colors, 149
   from people, 140
isotactic, 69, 181
   polypropylene, 69
   polystyrene, 73
isotope separation, 71

jam, 97
   cyanide in, 127
Japanese peppermint, 116
Jarvik heart, 82
jasmine, odor of, 133
Jasminum, 133
Javanese citronella grass, 136
juniper, odor of, 137

ketone, 58, 181
Kevlar, 87
kidney, 168
kieselguhr, 172
Kipper brown, 45
kippers, 45
kneading, 101
knocking, 38
kola, 169
Krakatoa, 20
krypton, 16

lachrymator, in onions, 131
   in smog, 32
   in smoke, 123
lactalbumin, 57
lactase, 99
lactic acid, 48
   and cramp, 101
   dextro form, 48
   levo form, 48
   margarine flavor, 59
   and odor of sweat, 140
Lactobacillus bifidus, 57
Lactobacillus bulgaricus, 49
Lactobacillus plantarum, 49
Lactobacillus sanfrancisco, 48

lactose, 100
lager, 113
lamb fat, softness of, 55
lamps, fluorescent, 16
   incandescent, 16
   indicator, 17
latex, 77
lauric acid, 62
Laurus nobilis, 135
lauryl methacrylate, 75
lead acetate, 92
lead azide, 172
leavening, 101
leavening agent, 22
leaves, color of, 150
   turning yellow, 151
lecithins, 59
legs, in nylon stockings, 86
   and wine, 52
legumes, and fixation, 19
lemonade, taste of, 111
leopard skin, 65
leucine, 89
Leuconostoc mesenteroides, 49
levulose, 97
Lewis, G. N., 5
Librium, 164
light, infrared, 181
   ultraviolet, 17, 182
   white, 149
lighter fuel, 36
lignin, 102
lignocellulose, 102
limbic system, 124, 164, 166, 181
   effect of stimulants on, 167
limestone, 21, 34
linalool, 133
line formula, 8
linoleic acid, 57, 58
linolenic acid, 55
linseed oil, 55
Linum usitissimum, 55
lipid, 54, 181
   metabolism, 56
liquid, 10
   ammonia solutions, 26
liquid air, 17
lobster, boiled, 154

color of, 154
locus coeruleus, 167
lone pair, 8, 181
longleaf pine, 137
LPG, 34, 36
Lucite, 73, 74
lunar rock, 17
luster, of hair, 92
luteolin, 157
lycopene, 152
Lycopersicon esculentum, 152

mace, 135
macrofibril, 91
magnetism, 181
Maillard reaction, 158
malaria, and quinine, 112
malt, 44
malting, 113
mango, color of, 151, 153
maple syrup, 98
   color on boiling, 158
margarine, 57
   color of, 57, 59
   soft, 59
marijuana, 170
Mars, atmosphere of, 21
marsh gas, 35
mash, 113
meat, color of, 117
   flavor of, 120
meatiness, 117
melanin, 31, 158
   in hair, 92
melon, nonbruising, 159
melting point, of chocolate, 139
menopause, 177
Mentha arvensis, 116
Mentha piperita, 116
Mentha viridis, 133
menthol, cool taste of, 116
Mephitis mephitis, 142
mercaptan, 142
mercury, 4
metabolism, 21
   of glucose, 96
metal ammonia solutions, 26
methamphetamine, 168

methanal, 45
methane, 2, 34
methanol, 42
methedrine, 168
2-methoxy-5-methyl pyrazine, 129
methyl alcohol, 42
methylacrylate, 74
methyl cyanoacrylate, 76
methyl methacrylate, 74
methyl 2-pyridyl ketone, 129
3-methylbutane-1-thiol, 142
2-methylpropene, 79
mica, 23
micrococcus, 121
microfibril, 90
microscopy, electron, 1
   surface tunneling, 4
migraine, 163
milk, butter production from, 57
   coagulation of, 57
   color of, 151
   curdled, 49
   drinking before alcohol, 54
   goat's, 58
   homogenized, 57
   human, 57
   indigestibility of, 100
   sheep's, 58
   translucency of, 68
milk sugar, 99
mini-pills, and progestogen, 177
mink, odor of, 142
mint, 116
mirror image, and odor, 134
mirror image molecules, and
   stimulating effect, 168
   taste of, 107
   and thalidomide, 174
mocha, 130
modacrylics, 83
Mogadon, 164
molasses, 97
molecular formula, 9
molecular solid, 11
molecule(s), 5
   atactic, 73, 180
   mirror image, and

stimulating effect, 168
  taste of, 102
  and thalidomide, 174
molybdenum, 19
monoglyceride, 52
monomer, 66, 181
monosaccharide, 98
monosodium glutamate (MSG),
  120
mood modification, 162, 167
morphine, 165
  action of, 165
  receptor, 165
Moschus moschiferus, 141
moth, night peacock, 93
  silk, 93
MSG, 120
multipass digestion, 102
muscle, and anabolic steroids,
  176
  fast, 117
  tenderness of, 117
muscle fiber, fast and slow,
  117
muscle relaxants, 164
muscle tissue, 117
mushrooms, and meaty taste,
  121
musk, 141
  of deer, 141
muskone, 141
mustard gas, 173
mycelium, 104
Mylar, 81
myocardium, 56
myoglobin, 117
myosin, 117
Myristica fragrans, 135

N-acetyl-*para*-aminophenol,
  163
nails, 92
natural gas, 35
nectar, 96, 97
neon, 16
  in advertising signs, 17
  indicator lamps, 17
neoprene, 79
nettle, stinging, 47

neurons, highly branched, 167
neurotransmitter, 43, 159
newsprint, 102
nicotine, 110
night peacock moth, 93
nitrates, 31
nitrazepam, 164
nitric acid, 30
nitric oxide, 30
nitro group, 171
nitrogen dioxide, 30
nitrogen fixation, 19, 26
nitrogenase, 19
nitroglycerin, 172
noble gas, 16
noble rot, 50
nodules, 20
nonstick coatings, 72
nonionic detergent, 64
noradrenaline, 167
norepinephrine, 167, 168
NOX, 30
nucleoside, 119
nucleus, 3
nudibranchs, 29
nutmeg, 135
nuts, color of, 158
nutty flavor, 46
nylon, 84
  and electrostatic charge, 86
  extrusion of, 86
  as insulator, 86
  stockings, 86
nylon-6, 84
nylon-6,6, 84, 86
nylon salt, 86

oak barrels, vanilla in, 138
oceans, 23
octane, 37
  number, 38
octanoic acid, formula of, 9
octopus, 159
odor, of almond, 127
  of ammonia, 26
  of animals, 140
  of armpit, 58, 140
  of barbecue, 123
  of bitter almonds, 126

blindness, 125
of bread, crusty, 130
of caramel, 123
of cheese, 26
  blue, 128
of cherries, 127
of chocolate, 130
of clove, 128, 135
of coffee, 130
of coyote, 144
of dairy products, 128
of death, 122, 144
of dogs, 144
of eggs, rotten, 122
and emotion, 124, 140
of feet, unwashed, 58
of ferret, 142
of fish, rotten, 144
of flatulence, 140
of flesh, rotting, 144
of food, 124
  heat-treated, 130
of fox, red, 142
of fruits, 128
  ripe, 46, 127
of garlic, 131
of geranium, 136
of hay, 129, 152
of hot rubber, 78
of hyacinth, 126
of hydrogen sulfide, 122
of jasmine, 133
of juniper, 137
mechanism of, 125
of mink, 142
of nutmeg, 135
of onion, 131
of peanuts, 130
of people, 78, 140
of pepper, 130
of popcorn, 130
of raspberries, 129
of rose, 136
of rum, 130
of semen, 144
of skunk, 142
of smoke, 123
of stoat, 142
of sweat, 140

of urine, 26, 144
of violets, 129
of whisky, 130
odorivector, 124
oil(s), 50
  of aniseed, 136
  of bay, 135
  of cinnamon, 134
  of citronella, 136
  of clove, 128, 135
  cod liver, 55
  corn, 55, 59
  drying, 55
  essential, 132
  fish, 55
  frying, and aging, 55
  fusel, 46
  of jasmine, 133
  olive, 55
  of pine, 137
  rapeseed, 54, 59
  safflower, 55
  salad, 55, 59
  of spearmint, 133
  sunflower, 55
  of turpentine, 137
  of vanilla, 138
  of violets, 129
  viscostatic, 76
  of wintergreen, 162
  winterized, 55
oil-based paint, 55
Olea europaea, 55
oleic acid, 54
olfaction, 106, 124
olfactory epithelium, 124
oligosaccharide, 99
olive oil, 55
onion, odor of, 131
opiates, 114
opium, 165
  poppy, 166
opsin, 147
optical activity, 49
oral contraceptives, 177
orange, blood, 12
  color of juice, 153
organic compounds, 6
organometallic compound, 39

Orlon, 83
osmophore, 124
osmotic pressure, 181
outgassing, 15
oxalic acid, 111
oxidation, 181
oxygen magnetism, 18
oyster mushrooms, 121
ozone, 22, 30
    and tanning, 160
    layer, 22

pain, 105, 114
    effect of morphine on, 166
    fast, 114
    signals, 163
    slow, 114
pain killers, 162
pain nerves, 114, 162
    types of, 114
paint, acrylic, 74
    alkyd, 55
    oil-based, 55
    white, 68
palmitic acid, 55
PAN, 32
pancreatic lipase, 56
Papaver rhoeas, 156
Papaver somniferum, 165, 166
paper pulp, 31
paprika, 115
para-hydroxyphenol-2-butanone,
    128
para-xylene, 81
paracetamol, 163
paraffin wax, 57
paramagnetism, 181
pea, raffinose in, 99
peach, cyanide in, 127
peanuts, odor of, 130
peat, 44
pelargonidin, 156
pelargonium, 156
Penicillium roquefortii, 128
peonidin, 156
peony, color of, 156
pepper, action of, 115
    aroma of, 130
    black and white, 115

peppermint, 116
peptide, 182
peptide group, 182
perfume, civet, 141
    and essential oils, 132
    fixative, 141
    hyacinth, 126
    musk, 141
permanent waving, 91
peroxide, 31
    and rancidity, 60
peroxyacetyl nitrate, 32
persimmon, color of, 151, 152
Perspex, 74
perspiration, acidity of, 49
    odor of, 58
pharmaceuticals, 50
phenol, 182
    in smoke, 123
phenol oxidase, 159
phenolic compounds, as
    antioxidants, 60
phenyl group, 73
phenylalanine, 109
phenylethanol, 135
pheromone, 140
phosphate, in detergents, 61
    fertilizer, 29
phosphodiesterase, 169
phosphor, 16
phosphorescence, 181
photochemical reaction, 182
photosynthesis, 17, 21, 150
phthalic acid, 82
pickles, fermented, 49
pigmentation, of skin, 159
pigs, color of, 18
    and truffles, 140
Pimpinella anisum, 136
pine oil, 137
pinene, 137
pink color, 154
Pinus elliottii, 137
Pinus palustris, 137
piperine, 115
plants, flowering, and the
    dinosaurs, 110
plasticizer, 50, 70
Pleurotus ostreatus, 121

Plexiglas, 74
poison, 171
poisoning, by carbon
    monoxide, 90
    by cyanide, 90
polarized light, 96, 182
    and sugar, 96
    rotation, 48
polarized light microscopy, 182
pollution, 27, 31
poly(ethylene terephthalate),
    81
poly(hexamethylene adipamide),
    85
poly(lauryl methacrylate), 76
poly(methyl cyanocrylate), 76
poly(methyl methacrylate), 74
poly(vinyl chloride) (PVC), 70
poly(vinylidene chloride), 70
polyacrylonitrile acrylics, 83
polyamide, 84, 89
polyester, 80
polyethylene, 67
    extrusion of, 68
    high-density, 68
    as insulator, 68
polyglycine, 89
polyisoprene, 78
polymer, 66, 182
polymerization, of frying oil,
    55
polyoxyethylene, 64
polypeptide, 88
polypeptides, 90
polypropylene, 69
Polyrhachis, 47
polysaccharide, 101, 102
polystyrene, 72
    color when aged, 73
    foamed, 73
polysynaptic neurons, 167
polytetrafluoroethylene, 71
polyunsaturated fat, 55
pome, 127
popcorn, odor of, 130
poppy, acidity, and color, 156
    color of, 3
    opium, 165
pork fat, softness of, 55

port, and gout, 49
Poriphera, 117
postsynaptic neuron, 43
poultry, color of meat, 118
poultry fat, softness of, 55
preservatives, 27
    in wood smoke, 45
preserved vertebrates, 51
presynaptic knob, 43
progestogen, 177
prokaryote, 182
proof, 44
propane, 36
propylene, 69
prostaglandin, 163
prostagladin cyclooxygenase,
    163
protein, 88, 182
proteolytic enzyme, 62
proton, 181
prowess, sense of, 168
pruritis, 115
PTFE, 71
Pulicaria dysenterica, 155
pulp, bleaching, 102
pungency, 125, 131
purines, 49
purple sulfur bacteria, 123
putrefaction, and amines, 85
putrescine, 144
putridity, 126
PVC, 69, 70
pyrite, 27

quercetin, 157
Quercus robur, 138
Quercus sessilis, 138
Quercus tinctoria, 157
quick-tanning lotion, 160
quince jam, cyanide in, 127
quinine, 112

rabbit, digestion in, 102
radicals, 39, 55, 182
    and flames, 83
    and rancidity, 60
radon, 16
raffinose, 99

rancidity, 60
rape plant, 54
rapeseed oil, 54, 59
raspberries, color of, 156
   odor of, 129
receptors for morphine, 165
   visual, 146
red color, in cabbage, 156
   in hair, 92, 159
   in Leicester cheese, 152
   in muscle fiber, 117
   in poppy, 156
Red Sea color, 20
reduction, 182
reforming, 38
renal gland, 168
retina, damage to, 45
   effect of methanol on, 42
   structure of, 146
retrogradation, 101
Rhizobium, 19
rhodopsin, 147
Rhodospirillum rubrum, 123
rhubarb, color of, 156
   toxicity of, 111
ribose, 119
rickets, 100
rigor mortis, 101, 119
ripe fruit, odor of, 46
   sweetness of, 96
ripening, of apples, 127
rock, 12
   carbonate, 21
   lunar, 17
   outgassing, 17
rods, in eye, 146
Roquefort cheese, 58
   odor of, 128
rose, odor of, 136
rosemary, 60
rouge, 55
rubber, butyl, 79
   odor of, 78
   SBR, 79
   stretching and recoil, 78
   styrene-butadiene, 79
rubber elastomer, 77
rum, aroma of, 130
ruminants, digestion in, 102

saccharin, 108
Saccharomyces cerevisiae, 22, 46, 101
Saccharomyces exiguus, 48
Saccharum officinarum, 98
safflower oil, 55
saffron, 152
saffron crocus, 145
sage, 60
St. John's wort, 155
sake, 96, 113
sal volatile, 26
salad oil, 55, 59
salicin, 162
salicylic acid, 162
   taste of, 110
Salix alba, 162
salmon, color of, 154
salt, 182
saltiness, 106
saltpeter, 31
sandstone, 34
sapidity, 107
sapodilla tree, 133
saporous unit, 107
saran, 71
satiety value, of food, 54
saturated compound, 182
saturated hydrocarbon, 53
sauce, thickening of, 101
sauerkraut, 49
SBR, 79
   as chewing gum, 134
scorpion, exoskeleton of, 104
Scotch whisky, 44
scum, 61, 63
sea slugs, 29
sea urchin, 97
seat covers, 71
sedative action, 44
sedatives, 164
seed, starch content of, 101
   oil content of, 55
seltzer, 21
semen, fructose in, 97
   odor of, 144
sense, of hot and cold, 114
sex, 175
shale, 34

shaving soap, 62
sheep's milk, odor of, 58
shells, 21
sherry, flavor of, 46
shortening, 59
shrimp, color of, 154
sight, 145
signs, advertising, 17
silicon, atoms in, 2
silk, 88, 93
   smoothness of, 93
   wild, 93
silk moth, 93
silkworm, 93
skin, color of, 159
   dye, 160
skunk, odor of, 142
slash pine, 137
slow muscle, 117
slow pain, 114, 166
smell, 105
   acrid, 123
   of animals, 140
   first and second, 126
   of food, 130
   sense of, 125
smog, 27
   color of, 30
   as lachrymator, 32
smoke, composition of, 123
   odor of, 123
   preservative action of, 45
   from TNT, 172
smoked fish, 45
smoothness, of glycerol, 50
   of silk, 93
soap, 61
   bubbles in, 33
   shaving, 62
soda water, 21
   sourness of, 110
sodium alkylbenzenesulfonate, 63
sodium bicarbonate, 22, 163
sodium para-dodecylbenzene sulfonate, 63
sodium stearate, 62
soft drinks, acidity of, 84
softness, 12

   of meat, 55
Solanum dulcamara, 110
solid, 10
   ionic, 11
   molecular, 11
sourdough, 48
sourness, 106, 110
spearmint, oil of, 133
spectroscopy, 1
speed, 168
sperm, and fructose, 97
spherulites, 67
spice(s), as antioxidants, 60
   and pain, 114
spider spinneret, 94
spider web, 93
spinach, taste of, 111
spinneret, 81, 86
   of spider, 94
spirits, 44
sponges, and collagen, 117
sport, and metabolism, 101
staling, of bread, 101
starch, 95, 101
   hydrogen bonding in, 101
starch-sugar, 96
stearic acid, 53
steelmaking, 16
stercobilin, 143
steroids, anabolic, 176
stimulant(s), 167
   in chocolate, 139
stinging caterpillar, 47
stinging nettle, 47
stoat, odor of, 142
stockings, 86
straight chain, 37
strawberries, color of, 156
Streptococcus albus, 58, 140
Streptococcus thermophilus, 49
Streptomyces, 97
strychnine, 110
styrene, 72
styrene-butadiene rubber, 79
   as chewing gum, 134
substantia gelatinosa, 114, 166
sucrose, 98
sugar, invert, 99
sugar beet, 98

sugar cane, 98
sulfonates, 63
sulfur, 4, 5, 27, 28
  formation of deposits, 123
sulfur dioxide, 27
sulfur trioxide, 27
sulfuric acid, 29
sunburn, 160
sunflower oil, 55
sunscreen, 160
  quick-tanning lotion, 160
suntan, 160
Super Glue, 76
surface tunneling microscopy, 4
surface-active agent, 61
surfactant, 61, 182
  in margarine, 59
sweat, acidity of, 49
  odor of, 58, 140
  underarm, 140
sweet wine, 173
sweeteners, aspartame, 109
  cyclamate, 109
  saccharin, 109
sweetness, 106
  of glycerol, 50
synapse, 43, 182
synaptic vesicle, 43
synergistic interaction, 44
syrup, 96
  corn, 96
  maple, 98

tallow, 62
tan, 158
  formation of, 160
tannin, 156
tarragon, 136
tartaric acid, and baking
  powder, 22
taste, 105
  of apples, 111
  of beets, 84
  of beryllium salts, 107
  blindness, 136
  of chocolate, 139
  of citrus fruits, 111
  cool, 116

of chocolate, 55
  of menthol, 116
  of food, 124
  of gin and tonic, 112
  of lemonade, 111
  of meat, 120
  of menthol, 116
  of mirror image molecules,
    107
  of mushrooms, 121
  sharp, 84
  of soft drinks, 84
taste bud, 106
taste enhancer, 21
tea, color of, 159
  stimulant in, 169
tears, 32
  and onions, 131
  produced by smoke, 123
Teflon, 72
teratogenicity, 174
terephthalic acid, 80
terpene, 132, 150
terpineol, 137
2-tert-butyl-4-methoxyphenol,
  60
terylene, 81
testes, 176
testosterone, 176
tetraethyllead, 39
tetrafluroethylene, 71
tetrahydrocannabinol, 170
thalamus, 166, 182
  role in pain, 114
thalidomide, 174
Theobroma cacao, 139
theobromine, 139, 169
thermal receptors, 114
thiol, 142
thiopropionaldehyde-S-oxide,
  131
thyme, 60
tire cords, 87
titanium dioxide,
  in paint, 68
TNT, 41, 171
Tolu balsam, 41
toluene, 40
  formula of, 9

Toluifera balsamum, 41
tomato, color of, 152
  and nonbrowning bruising,
    159
tongue, 106
toothpaste, glycerol in, 50
top and bottom fermentation,
  113
tranquilizer, 162
  action of, 44
trans, 10
trans-retinal, 148
translucency, 68
  of ice, 68
transmutation, 182
transparency, of Lucite, 74
  of polystyrene, 73
  of water, 74
Trevira, 81
trichosiderin, 92, 159
trigeminal nerve, 126
triglyceride, 52
trimethylamine, 144
trimethylpentane, 38
trinitrotoluene, 171
tristearin, 53
  and soap, 62
tropocollagen, 117
truffles, 140
Tuberales, 140
Turacos, 143
turpentine, oil of, 137
Tylenol, 163
tyrosine, 159, 168

ultraviolet light, 17, 182
Umbellularia californica, 135
unsaturated compound, 182
unsaturated hydrocarbon, 55
uranium hexafluoride, 71
urea, 6, 143
urea-formaldehyde resins, 45
uric acid, 49
urination, 44
urine, 49
  cloudiness in, 143
  color of, 143
  human, 143
  odor of, 26, 142, 144

and uropophyrin, 143
Urtica, 47

vagina, acidity of, 49
Valium, 164
  action of, 44
vanilla, oil of, 138
vanilla orchid, 138
Vanilla fragrans, 138
vanillin, 8, 137
vegetarian urine, 143
vegetation, color of, 150
Velcro, 87
venom, 47
Venus, atmosphere of, 21
Venus and Mars, 177
vertebrates, preserved in
  glycerol, 51
vinegar, 48
  sourness of, 110
vinyl chloride, 69
vinylidene chloride, 70
Viola odorata, 129
violets, odor of, 129
viscostatic engine oil, 76
vision, 146
visual purple, 147
vitamin A, 150
vitamin D, 100
vitamin E, as antioxidant, 60
Viverra civetta, 141
vodka, 44
volcanoes, 27
vulcanite, 79
vulcanization, 77

wakefulness, 164
Warhol, Andy, 75
water, 23
  anomalous density of, 23
  as solvent, 24
  color of, 24
  hard, 63
  transparency of, 74
water-resistance, 87
wavelength, 182
web, spider, 93
wheat flour, 101
whisky, 44

aroma of, 130
white light, 149
white muscle fiber, 117
white willow, 162
white skin color, 18
wigs, 83
wild silk, 93
willow, white, 162
wine, 44, 155
   and aging in oak, 138
   and Austrian scandal, 173
   bouquet of, 42
   color of, 156

glycerol in, 173
legs in, 52
noble rot and, 50
oxidation of, 48
smoothness of, 50
sweet, 173
sweetness in, 50
winemaking, 27
wintergreen, oil of, 162
winterized oil, 55
wood, distillation of, 41
   as foamed polystyrene
      analogue, 95

structure of, 102
wood smoke, 42, 45
   odor of, 123
woody nightshade, 110
wool, 88, 90
   treatment of, 45
wort, 113
wrapping, cling, 71

x-ray diffraction, 1
xanthophyll, 151, 153
xenon, 16

xylene, 41
xylol, 41

yeast, 22
yellow colors, 157
yogurt, 49
Ypres, mustard gas at, 173

Zea mays, 153
zeaxanthin, 153
Zinfandel grapes, 52
zingerone, 116
Zingiber officinale, 116

OTHER BOOKS IN THE SCIENTIFIC AMERICAN LIBRARY SERIES

POWERS OF TEN
by Philip and Phylis Morrison and the Office of Charles and Ray Eames

HUMAN DIVERSITY
by Richard Lewontin

THE DISCOVERY OF SUBATOMIC PARTICLES
by Steven Weinberg

THE SCIENCE OF MUSICAL SOUND
by John R. Pierce

FOSSILS AND THE HISTORY OF LIFE
by George Gaylord Simpson

THE SOLAR SYSTEM
by Roman Smoluchowski

ON SIZE AND LIFE
by Thomas A. McMahon and John Tyler Bonner

PERCEPTION
by Irvin Rock

CONSTRUCTING THE UNIVERSE
by David Layzer

THE SECOND LAW
by P. W. Atkins

THE LIVING CELL, VOLUMES I AND II
by Christian de Duve

MATHEMATICS AND OPTIMAL FORM
by Stefan Hildebrandt and Anthony Tromba

FIRE
by John W. Lyons

SUN AND EARTH
by Herbert Friedman

EINSTEIN'S LEGACY
by Julian Schwinger

ISLANDS
by H. William Menard

DRUGS AND THE BRAIN
by Solomon H. Snyder

TIMING OF BIOLOGICAL CLOCKS
by Arthur T. Winfree

EXTINCTION
by Steven M. Stanley